REGENERATIVE
LEARNING

Regenerative Learning is a ground- breaking and urgent call to abandon the traditional evolutionary perspective which places humans at the peak of the ecological tree to exploit and despoil everything below.

A new ethical and environmental education can teach us the responsibility and stewardship that we need if we are to have any possibility of continuing, with nature, to flourish.

Published for the 30th anniversary of Schumacher College, UK.

Satish Kumar is the author of many books (most recently, Elegant Simplicity). He is also Editor Emeritus of Resurgence & Ecologist, and Founder of The Small School as well as of Schumacher College, U.K.

Lorna Howarth is a writer, editor and publisher. For more information visit www.lornahowarth.com

REGENERATIVE
LEARNING

Nurturing People and Caring for the Planet

Edited by
SATISH KUMAR & LORNA HOWARTH

GLOBAL
RESILIENCE
PUBLISHING

GLOBAL
RESILIENCE
PUBLISHING

An imprint of
Salt Desert Media Group Limited,
7 Mulgrave Chambers, 26 Mulgrave Rd,
Sutton SM2 6LE, England, UK.
Email: publisher@saltdesertmedia.com
Website: www.globalresiliencepub.com

ISBN 978-1-913738-46-4

Designed and typeset by Raghav Khattar

Printed and bound at Replika Press, Sonipat

*In celebration of the 30th anniversary of
Schumacher College, this book is dedicated to all
the staff, teachers, students, and other-than-human
beings who have made Schumacher College such an
adventurous and joyful learning environment.*

Contents

To celebrate 30 years of pioneering education at Schumacher College, we organised an essay competition, 'Regenerative Learning: Education as if People and Planet Matter.' All entries were judged by an independent panel.

Exploring the history of education, why we came to develop the educational systems we currently adhere to, and how we can design diverse and regenerative educational systems for a vibrant and harmonious future.

PART THREE   187
In Practice

Examining how we can make the structural changes we so desperately need, in order to lead us towards a regenerative education which prioritises not money and narrowly-economic growth but people and the planet.

Breathe Out

Courageously Pursuing New Educational Horizons

WITH GRATITUDE TO ALL TEACHERS AND CONCERN FOR EDUCATION

H.H. Pope Francis

In September 2019, I appealed to all those engaged in the field of education to "dialogue on how we are shaping the future of our planet and the need to employ the talents of all, since all change requires an educational process aimed at developing a new universal solidarity and a more welcoming society." For this reason, I promoted the initiative of a 'Global Compact on Education' in order to "rekindle our dedication for and with young people, renewing our passion for a more open and inclusive education, including patient listening, constructive dialogue and better mutual understanding." I invited everyone "to unite our efforts in a broad educational alliance, to form mature individuals capable of overcoming division and antagonism, and to restore the fabric of relationships for the sake of a more fraternal humanity."

If we desire a more fraternal world, we need to educate young people to acknowledge, appreciate and love each person, regardless of physical proximity, and regardless of where he or she was born or lives. The fundamental principle 'Know yourself' has always guided education, yet we should not overlook other essential principles:

- 'Know your brother or sister,' in order to educate in the welcoming of others;
- 'Know creation,' in order to educate in caring for our common home; and
- 'Know the Transcendent,' in order to educate in the great mysteries of life.

We cannot fail to speak to young people about the truths that give meaning to life.

Religious traditions have always had a close relationship with education and, as in the past, so also in our day, with the wisdom and humanity of our religious traditions, we want to be a stimulus for a renewed educational activity that can advance universal fraternity in our world. If, in the past, our differences set us at odds, nowadays we see in them the richness of different ways of coming to God and of educating young people for peaceful coexistence in mutual respect.

If, in the past, also in the name of religion, discrimination was practiced against ethnic, cultural, political and other minorities, today we want to be defenders of the identity and dignity of every individual, and to teach young people to accept everyone without discrimination. For this reason, education commits us to accept people as they are, not how we want them to be, without judging or condemning anyone.

If, in the past, the rights of women, children and the most vulnerable were not always respected, today we are committed firmly to defend those rights, and to teach young people to be a voice for the voiceless. Education must help us realise that men and women are equal in dignity; there is no room for discrimination.

If, in the past, we tolerated the exploitation and plundering of our common home, today, with greater awareness of our role as stewards of creation, we want to give voice to the plea of Nature for its survival, and to train ourselves and future generations in a more sober and ecologically sustainable lifestyle. When I heard one of the scientists at our meeting say, "My new-born granddaughter will have to live, in fifty years' time, in an unliveable world, if things continue as they are," I knew that education must commit us to love our Mother Earth, to avoid the waste of food and resources, and to share more generously the goods that God has given us for the life of everyone.

And so, I invite you to courageously pursue the paths of new educational horizons.

Extract from a speech given by H.H. Pope Francis given on UNESCO's World Teachers' Day, at the Global Compact on Education, October 2021. Published here by permission.

Introduction

EDUCATION THAT IS FIT FOR THE FUTURE

REGENERATIVE EDUCATION IS THE BEATING HEART OF VIBRANT, HEALTHY, AND JOYFUL COMMUNITIES

Satish Kumar and Lorna Howarth

To celebrate the 30th Anniversary of Schumacher College, in association with *Resurgence & Ecologist*, we launched an essay competition on the subject, 'Regenerative Learning: Education as if People and Planet Matter'. We wanted to explore current thinking about how education can and must evolve in order to rectify the many shortcomings it now embodies. Much to our delight, we were inundated with entries for the essay competition – almost 250, in fact.

An independent panel of judges, chaired by veteran environmentalist Jonathon Porritt, chose the prize-winners. Although there could only be three winning essays – which are of course featured in this book – many of the other essays made such good reading that we decided to publish them as a collection. We also commissioned a few more contributions from leading thinkers and activists in the field of regenerative education to make what we think is a fascinating 'bouquet' of essays on the subject.

It has been a privilege to read these contributions – not just from educators but scientists, artists, gardeners, musicians, and young people themselves – and to feel a powerful consensus arising: the present systems of education are fundamentally flawed and not fit for purpose. This is hardly surprising, because the dominant global educational system was designed to meet the needs of the Industrial Age, the age of mass production, mass consumption and unlimited economic growth, where young people were trained in whatever skills the market required. This was an education-for-jobs rather than an education-for-life. Many of the jobs for which students were trained, and are still trained, have led to the multifaceted and catastrophic tipping points which society now faces. In fact, all our myriad contemporary crises have been precipitated by 'educated experts', many of whom went to the world's best schools and universities. We have to ask ourselves: why are we still giving the kind of education that has brought us to the precipice of disaster?

Fortunately, awareness is growing that 'what we do to the Earth, we do to ourselves'. This awareness has brought us to the threshold of a new era, which Stephan Harding calls, 'The Age of Enchantment' where the rights of the Earth are equal to those of human rights. The coming of such a time cannot happen soon enough, and it requires imaginative, learner-led, experiential and holistic systems of education, which are Nature-centred, and can help to develop a truly regenerative and resilient culture.

In November 2021, when most of the world's political leaders met at the COP26 summit in Glasgow, UK, they made promises to halt carbon emissions and safeguard biodiversity and environmental integrity. They spoke about changes in government policies, changes in business practices, and a need for new technologies to address the urgency of climate change. These express good sentiments, but everyone knows that sentiments are hardly enough. It seems that politicians around the world are unwilling to face the fact that current systems of education are not equipping our young people with the skills they need to fully participate in maintaining the integrity of our precious planet Earth and the wellbeing of humanity. In that sense, COP26 was a shameful failure.

The articles compiled in this book suggest powerful ways in which we can begin to make real progress. Such progress must start with a root and branch rethinking of education. Unless we change our educational system and learn a new way of living in harmony with each other as well as the natural world, the future is bleak indeed. The good news is that we do not have to reinvent the wheel. For years there have been many examples which have acted like beacons of light across the educational establishment that have signalled another way of learning, where the classroom is extended out into Nature, so that we can dialogue with other-than-human realms.

Imagine a system of schooling where, for example, beekeeping teaches us about community; boat-building teaches us about trees and physics; star-gazing teaches us about our ancestors' navigational excellence; rock-climbing teaches us about the qualities of balance and focus; gardening teaches us about nutrition, doughnut economics and patience; where art connects us with our infinite imagination; and where music matches with maths. When Nature is our classroom, all things converge, interconnect, and make sense.

What is needed to make education relevant for people and planet is a change of perspective: from "education for dominance and conquest" to "education for stewardship" – in other words, "education of participation and imagination". Changing the perspective from top-down to bottom-up also allows education to be student-led, where the educator facilitates the knowledge students want to learn, not merely what they are forced to learn. When we do this, magic happens, and magic can happen in classrooms all over the world if we let go of our obsession with tests and exams and allow students to discover the miracle of Nature and mystery of human imagination.

In his article, 'A School of Hope,' Gunter Pauli and his colleagues tell the story of a school in Croatia that reinvented itself after the Balkan War; where educators and innovators listened to the students, the teachers, the parents and the wider community, and helped them to create the kind of school that the future demands: one that generates its own electricity, that grows much of its food, that apprentices students in the community and the community in the school, and that listens – really listens – to what the young people are saying they need. When all these spheres come into resonance, what results is a synergy that gives children their future back.

We are delighted to offer this collection of radical and profound essays to all parents, teachers, lecturers, professors, and policy makers, in the hope that they will be inspired to transform educational systems and make them fit for the future.

Satish Kumar is the author of many books (most recently, Elegant Simplicity*). He is also Editor Emeritus of* Resurgence & Ecologist, *and Founder of The Small School as well as of Schumacher College, U.K.*

Lorna Howarth is a writer, editor and publisher. For more information visit: www.lornahowarth.com

PART ONE

THE PRIZE-WINNING ESSAYS

First Prize

To Value a Wasp

WE CANNOT CONTINUE TO DIVORCE EDUCATION FROM LIFE

Matt Carmichael

*"Humanity will be saved not by more information,
but by more appreciation."*
– Rabbi Heschel

Wasps are welcome to disrupt my English classes. It's happened more often this spring because of the requirement to keep windows open during the pandemic. I once resented the working time lost coaxing the little gate-crashers back out of the window, students screeching, swatting, pronouncing the death sentence. But six years ago, something sticky-sweet on my finger from lunch presented me with an opportunity. I let the little creature linger on my open hand. "Watch out, sir! They can sting without dying, sir, not like bees!"

"Why would it sting me? In late summer they can be grumpy but this one's fine. I think it's a male because its abdomen's quite slim."

I'm among the desks, and a few students lean closer to see. "You know, without wasps, we'd be plagued by swarms of flies and midges. Wasps keep our ecosystems in balance. Without wasps, our food would be more expensive and less healthy because farmers would use more toxic spray to protect crops. There are 30,000 species of wasps and most are pollinators just like bees. There are even some beautiful orchids that wouldn't exist without wasps; they've evolved to look and smell like female wasps to trick males into pollinating them. Did you know there are honey wasps?"

Students further away stand up for a better view. Some move closer. "In Japan you can eat wasp larvae in fancy restaurants." Disgusted cries. "There's research going on into a wasp venom that might one day save your life, because it kills cancer cells without harming healthy cells." A minute ago, we were in rows facing the front. Now we are in a circle.

In *Small Is Beautiful: A Study of Economics as if People Mattered*, the economist E.F. Schumacher identifies the heart of the problem: "We must decide what we want our economies to do for us, otherwise the relationship is easily inverted, and economies enslave us." He devotes a whole chapter to education, writing, "If western civilisation is in a state of permanent crisis, it is not far-fetched to suggest that there may be something wrong with its education." That crisis is today so glaring that many young people themselves are aware of the inadequacy of their own education system. Through the brilliant 'Teach The Future' campaign some are demanding a curriculum review so that sustainability and climate change are taught in all subject areas. How would that look?

I've devoted much of the past two decades to communicating the reality of climate change in Leeds to all kinds of audiences, and I've learned that people have a great appetite for scientific explanations, but that knowledge on its own yields unpredictable responses. Depending on their prior value systems, people I've addressed have done everything from disrupting oil-sponsored art exhibitions to starting a climate denial blog. So, education as if people and planet matter cannot rely on accurate information transfer alone. Indeed, if Schumacher is to be believed, we must address not just the content of the curriculum, but the purpose of life: "…'know-how' is nothing by itself," he writes, "it is a means without an end… the task of education," must be "first and foremost the transmission of ideas of value, of what to do with our lives."

This question cannot be avoided. Every education system is necessarily an expression of – and therefore a teacher of – particular values. Ours grew out of the liberal tradition – 'liberal' because it aimed to free thinking from stifling religious dogma. Its curriculum, introducing the subjects we still learn, served the industrialising economy and nascent democracy, but made assumptions which are at the root of the crises we now face: that Nature is a soulless mechanism to be moulded to human purposes; that its stock is effectively limitless; and that waste products are of no great significance.

It's striking that the rise of neoliberalism – a new stifling dogma sanctifying the freedom of markets to generate profit – coincides roughly with destruction on a vast new scale, where 80% of all carbon has been emitted, and 60% of total animal populations have been lost, and striking too that the education system has been refashioned to enshrine neoliberal values. Headteachers are trained as managers of learning factories which compete with each other for tomorrow's children by proving that they are filling today's children up – like milk bottles on a conveyor belt – with the knowledge and skills to compete with each other in a neoliberal jobs market.

Education as if people and planet matter must be based on very different, life-affirming values. Most teachers are motivated by a desire to care for children, so even now, in small ways, the system subverts itself covertly. In fact, I think this is what happens when a wasp disrupts my English class. Years after that first wasp lesson, a student gave me a thank-you card when she left school. She said that I had inspired her to participate in the Youth Strike for Climate and to study philosophy at university. To my astonishment, she cited not my many carefully planned lessons and assemblies around climate change, but "when you held the wasp."

> Most teachers are motivated by a desire
> to care for children, so even now, in small ways,
> the system subverts itself covertly.

Pondering this has led me to the view that education's purpose is best expressed not just in terms of Schumacher's abstract ideas of value, but in terms of concrete relationships. Answering his unavoidable question places people in relationship with each other and Nature, and these roles become the fundamental learning outcomes of the education system, feeding through into the economy and wider society. When I held the wasp, students' relationships with wasps were transformed from something like antagonism to something more like allyship, and, for at least one, this set her on a new path.

There are several dimensions to the way this transformation takes place in the wasp lessons. Firstly, there are some facts which students have to make sense of. They are about how wasp and human interests align. The eco-philosopher Freya Mathews says, "if my identity is logically interconnected with the identity of other beings, then... my chances of self-realisation depend on the existence of those beings... our interests converge." Such information is necessary if people and planet both matter.

Secondly, an emotional dimension moves students from alarm towards empathy. In *Life's Philosophy*, the philosopher of deep ecology Arne Naess argues for an education that takes, "more account of feelings," committing a chapter to cultivating 'A feeling for all living beings.' An education system fit for the future today's children face would produce emotionally literate young adults, more aware of deep motives in themselves and others, and experienced in conflict resolution.

But without a wasp present in the room, I doubt I'd have received that thank-you card - just as, reading this essay, you cannot ask some of the questions my students asked: "What's it doing now?" "Can I hold it?" The presence of a living being is compelling, but factory education is addicted to smartboards, as though consciously acclimatising children for a semi-virtual life. What if children's curricular entitlement was expressed in living encounters rather than topics? Outdoor classes would be a daily expectation. Visits by artists, asylum seekers, and war veterans would be as commonplace as textbooks. Neoliberal education relies on divorcing school life from community life, but students of every age should be deeply involved in serving their local communities, for example by growing food, visiting old people, and creating what Schumacher called intermediate technologies.

Finally, there is magic in the spontaneity of the wasp lesson. Naess links the central "feeling of being on a voyage of discovery," to slower, deeper learning within a spacious curriculum. Factory learning is ruled by the monstrous god Chronos, but wise education must revere the Greeks' friendlier god of time, Kairos, whose educational incarnation is the teachable moment. This spring, when he has alighted incarnate as a wasp, I've dived into big questions: Given that there is no known biological life anywhere else in the universe, how valuable is a wasp? Who gave you the idea that it's ok to kill wasps? If 'everyone' thinks so, does that make it true? (So, the best of liberalism still contributes!) What kind of education system tells you about subjunctive clauses before you're twelve, but never explains why we need wasps? What else is it not telling you?

2,400 years ago, the Taoist teacher Zhuang Zhou wrote, "I know the joy of the fishes in the river through my own joy as I go walking along the same river." If we take young people for those walks, literally and figuratively, the relationships they form with people and planet will empower them to work out the rest for themselves. A student recently said: "Sir, I think you've convinced my brain that wasps are okay, I just don't know if I like this one." Before I could respond, someone else piped up, "Sammy. He's called Sammy. Bet you don't want to kill him now!"

Matt Carmichael has taught at an inner-city high school for over 20 years. He has long worked to achieve city level responses to the climate and ecological emergency, more recently focusing especially on schools. He lives in Leeds with his wife Kath and primary school aged children, Finlay and Lola.

Second Prize

WE ARE TREES

FOSTERING AN EDUCATION THAT ENABLES US TO BE TRULY AT HOME ON THE EARTH

Dheepa R. Maturi

"Teaching children about the natural world
is one of the most important events in their lives."
– Thomas Berry

The little creature was lopsided, scrambling on the ground, its leg obviously broken. My hand smudged the window as I watched my son push aside the mulch with his fingers and gently scoop the frightened bird into a box he'd pierced with air holes and filled with soft stuffing. As he drove off to a bird sanctuary, I thought about him and his brother as young children, their fingers pushing mulch, clutching grass, grasping branches. I thought of their bare feet running across dirt, shuffling through leaf piles, splashing in puddles. I remembered two small figures jumping up and down, sticks in hand, shouting, "Look — we are trees! We are trees!" And I remembered myself, too, armed with wet wipes and antibacterial gel, holding off as long as I could before swooping in to clean those fingers and feet. I remembered telling myself not to worry, that they could have an extra bath later.

Looking back now, I am glad that I was able to sit on my obsessive-compulsive tendencies, at least temporarily, and that I had the sense not to interrupt the communion happening in front of me. I am thinking about all of my separations — how I watch life happening through windows and the world unfolding through the screens of modern technology. Inherently, I am a homebody, defining that home as the four walls around me. I have an aversion to dirt and bugs, a preference for wearing shoes and sitting in chairs. I am realising how disconnected I am from my cultural inheritance, specifically, from the deep respect for the soil beneath me. As a six-year-old, I watched a graceful lady impart my first classical dance lesson: bending at the knees and touching the ground in apology for the upcoming striking of my feet. The morning mantra I heard from family members expressed a similar apology before rising from bed and placing even one toe on the ground. The pre-dining mantra intoned profound gratitude to Mother Earth for being the source of not only the food, but also the people harvesting and preparing it.

Embedded within that same cultural mechanism was continuous contact with the ground, all day long: walking barefoot, sitting cross-legged, bending into the geometries of yoga. Modern science confirms that repeated and direct physical contact with the Earth's endless supply of electrons has a grounding effect, contributing to the bioelectrical stability of bodily systems. But from a psychological or even spiritual point of view, I suspect that all these actions also breed a constant consciousness of the Earth under one's feet and expand one's sphere of awareness to include that Earth.

I regret that so much ancient wisdom like this has been lost, kneaded away by time, leavened by globalism, and flattened by appropriation. So much has been rejected by those born to it, due to the self-doubt and self-hatred deliberately cultivated by colonialism and now enduring as a pernicious legacy. I regret that such influences amount to yet another separation I have allowed within my life, despite my best efforts. Thankfully, however, this ancient wisdom must still be coded within my genetic and karmic structures, because it has leapfrogged and manifested within my children. In spite of my shortcomings, they have received their inheritance, hands open, and embraced it wholeheartedly. I recognise that, within our relationship, the movement of wisdom between generations is not unidirectional. Indeed, from them I have learned lessons of greater and more abiding value.

> Ancient wisdom must still be coded within my genetic and karmic structures, because it has leapfrogged and manifested within my children

After all, it was their desires, their inclinations, that pulled me into city parks when they were young, and then, as they grew older, into the foliage of state forests and the rock formations of national monuments. They are the ones who made me place my own fingers and feet on the planet. They are the example I am trying to follow, as their free time is never spent within four walls, but rather, hiking on trails, climbing over rocks, camping under stars. To them, air means the mountain breeze and water, the ocean waves. Vacation means days spent walking on and breathing in the world around them. The difference between us is simply this: their definition of home is far richer and wider than mine. I am at home in my house, and they are at home on the Earth. There are no separations for them, only inclusions.

They are science and math types, these boys. They are typical young people with phones and screens, just like their peers. At work and at school, they analyse and dissect, label and programme. When I observe them, though, I see roots growing, sprawling in all directions. I see them as part of a system, their fibres and branches intertwining with place and people. All of their analysis and dissection, their labelling and programming, happens within this rootedness.

Their wide definition of home has made them unable to learn and act within a vacuum, unable to separate their education about the world, from the world. As such, they engineer with consciousness of the land on which structures are built, with consciousness of the people for whom they build. Home is the wide Earth, and by extension, its people are family, and all are tightly interwoven.

I surmise that education as if people and planet matter requires certain kinds of expansion. It means pushing lectures and readings outward into tactile interactions with land and community. It means pulling the definition of home outward, so that it becomes the foundation and context for every lesson, from mathematics to computer programming. It means demonstrating how every theoretical concept touches the world in some way.

For some years, I directed an education grant programme for teachers dedicated to adding context to curricula. With this financial support, biology students learned about the dwindling population of monarch butterflies by planting a pollinator garden in a local park. Elementary school students learned the dangers of invasive species by conducting mass weeding and by lobbying the Department of Natural Resources to protect native plants. The programme's goal was to show students that their education was intimately connected to life happening around them. While doing this work, however, my biggest surprise was the students' lack of surprise. This was not a revolutionary concept for them — they already knew the lesson, deep in their bones. It occurs to me that we adults thwart children from their inborn understanding, that we bully them into forgetting. We pull their natural comprehension of home and family inward, shrinking it down from planet to country to household, and into even smaller and smaller units.

Those children are breaking free, and not a moment too soon. So many of the next generation are determined to hold onto their birthright of clean Earth and connected community. So many have become warriors for this cause, planting trees to rebuild forests, developing technologies to filter oceans, lobbying governments to change laws. They are determined to preserve the ancient knowledge that every living thing, large and small, is intertwined with the next. They are refusing separation.

I think of that little broken bird often, lurching and faltering on the ground. I think of humanity lurching and faltering on a planet it has taken to the brink of destruction. I think of the struggles of modern life, with its broken support systems, broken communities, broken connections. Then, my hope returns. I think of the next generation: one that puts its hand to the Earth, feels a beating heart, and does the needful.

Dheepa R. Maturi writes to celebrate Nature and generate hope during this time of climate grief. A graduate of the University of Michigan and the University of Chicago, her essays and poetry have appeared in Literary Hub, Fourth River, New York Quarterly, Tweetspeak, Entropy, PANK, Dear America, The Indianapolis Review, *and elsewhere.* www.DheepaRMaturi.com

Third Prize

THE CIRCLE OF LIFE

LISTENING TO COUSIN BANANA, LEARNING FROM THE WOLVES

Guy Dauncey

"Let Nature be your teacher."
– William Wordsworth

My name is Isabella, and I am fourteen years old. I was born in 2107. I live with my parents in our village just outside the small town of Siquirres, in Limón Province, central Costa Rica. My *abuela*, my grandmother, has told me about the crisis that swept the world when she was young – how the forests were burning, how whole villages were swept away in floods, how the crops would fail because of the drought and the scorching heat; of how she used to love to go to the beach to swim in the sea, and watch the sea turtles – but now they're all gone. She would tell me of refugees from Nicaragua and Panama who came knocking on her door, children in their arms, begging for help.

But then she would pause and pat me on the knee. "But it's your world now, mi cielo. Tu nueva civilización ecológica. So, tell me about it. What are you doing at school? Do you enjoy it?"

So let me tell you about my school. My teachers are my heroes. We study the history of our country, and the role my grandmother's generation played in the political revolutions that changed Central America. We study Gran Historia – the incredible story of how our Universe came to be. We have learnt how every living thing has a shared family origin. We really are one. I share 93% of my genes with the spider monkeys, 50% with my cousins the bananas.

We start every morning with a circle where we talk about what's on our minds and in our hearts. Sometimes it's fun. Sometimes it's hard. Last week my friend Luciana told how her parents were splitting up, and how sad she felt, and how angry she was at her father. We held her in a circle. Most of us were crying. I love our morning circles. We talk about everything, including bullying, anger, hurt, resentment, jealousy, and the different kinds of love. Sometimes I leave the circle feeling confused, but mostly, my heart is singing. I feel loved. I love the students in my class.

In these circles, our teacher often reminds us of the North American Indigenous story about the two wolves who live within each of us, the one who wants to dominate and often gets angry, and the one who wants to be kind, and not hurt people. We share our stories of when we have behaved like one of the two wolves, and she tells us stories from history, encouraging us to learn about all the wars and brutalities and atrocities that have been caused by the wolf who wanted to dominate, and the incredible progress that has been achieved for ordinary people and for Nature by the wolf who wanted to be kind and cooperate. It makes me proud to be a human, to be part of the Circle of Life, and excited to be at the beginning of my great adventure.

Once a week we hold a different kind of circle – a Spirit Circle. We spend time sitting in silence, and then we share what we are sensing or feeling. We learn about the world's many religions, how much they have in common, and why they too caused people to fight and kill each other. We talk about our dreams, and the shape of our inner worlds, as our teacher Cora likes to call them. Last year my class went on a field trip to the Talamanca region where we got to sit with an elder from the Bribri people. Her people have lived in Costa Rica since time immemorial. She told us some of their stories, and their belief that all of Nature is sacred, the waters are sacred, the mountains are sacred, the birds are sacred. I think this is something I believe, as well.

So let me tell you more about our school. It's a fifteen-minute walk away, in Siquirres. It has ten circular buildings, each with a solar roof, arranged in a circle around a plaza where we gather and play. We've got a forest with trails, and tree platforms where we hold some of our classes. We have our own small farm where we grow vegetables and raise chickens, and we look after two elderly horses, who let us ride them. We have a big pond where we are supposed to be very quiet, because there are so many animals, birds and insects who live there. We have an amazing play zone that's totally wild, full of bits of wood and scrap things that we use to build things, and to hold mock battles. My father says it's dangerous, but my teacher tells him not to worry, for how can we learn about life if we are not exposed to danger, and risk?

Nature is a big part of our schooling. We have learnt how bad things were in the time of our grandparents, and how ignorant most people were about the climate, the forest, the ocean, and all the creatures we share life with. If you don't pass the exam in Ecology you can't go to university, and people frown on you. My sister had to take it three times before she passed. I think it's because she spent so much time on her computer, and not enough in the forest.

I wish that every child in the world
could have a school as good as ours

As well as the things we study following our natural curiosity, helped by our teachers, we are learning all sorts of practical skills. Earlier this year I spent five afternoons helping in the animal shelter in Siquirres, where they bring animals who are hurt or who have lost their mothers. I have learnt how to cut hair, how to cook, how to fix an electric bicycle, how to write code, how to plant trees, how to help install a solar panel, and how to tell a story to my *abuela* in a way that will make her smile.

I must tell you about our art, and our music! We have a choir that I sing in, and a school orchestra. I play the oboe. Once a month we do a bus trip when we visit the

Parc Nacional Barbilla, or go to San José, where we visit an art gallery, or the Museum of Civilization, or go to a concert by the National Symphony Orchestra. And once a year our choir gets together with choirs from all the other local schools, and we give a concert to our parents and all their friends. Last year there were more than a thousand of us, all singing together in harmony, accompanied by an incredible band with drums, guitars, violins, everything. It was absolutely wild. I couldn't sleep all night afterwards, because the music was pouring through my head.

What else? There's so much. We have an annual school election for various positions of leadership, and when I'm 16 I'll be able to vote in our local and national elections. We learn all about democracy, and the ways rich people have tried to take it away so that they can rule without bothering about ordinary people, farmers and workers like my family, and the people in our village.

This term I've chosen to study economics, because my *abuela* says it's important. I have been studying the economics of our local cooperatives. I spent a day helping at the Siquirres Community Bank, making coffee and cleaning dishes but also being allowed to sit in on meetings when people came in to discuss a loan. I have learnt how our country's trade laws changed when we ended the old colonialism, how we now favour trade that is fair by charging tariffs on imports from countries that have weak labour laws, or that don't do enough to protect Nature. I may be only 14, but I have been able to watch as one of our trade delegates negotiated a deal. Afterwards she came online, and she allowed us – children from all over Costa Rica – to ask her questions.

Do I have any complaints? Why would I have complaints? If I do, I can take them to our morning circle, or our democratic assembly. There are some students I'd like to complain about, but they'd probably like to complain about me. When I grow up, I want to be an economist, to help towns like Siquirres build their community wealth. Do I have a wish? Yes – I wish that every child in the world could have a school as good as ours.

This fictional piece is by Guy Dauncey, an ecotopian futurist who works to develop a positive vision of a sustainable future, and to translate that vision into action. He is the author of Journey to the Future: A Better World is Possible, *and co-founder of the West Coast Climate Action Network. He lives on Vancouver Island, in Canada.*

PART TWO

IN PRINCIPLE

Breathe In

A Good Model for Universities

I DO NOT UNDERSTAND HOW A POEM
CAN BE BETTER THAN A PEPPERMINT PLANT

Thich Nhat Hanh

I don't know what job you do every day, but I do know
that some tasks lend themselves to awareness more
easily than others. Writing, for example, is difficult to
do mindfully. I have now reached the point when I
know that a sentence is finished, but while writing the
sentence, even now, I sometimes forget. That is why I
have been doing more manual work and less writing
these past few years. Someone said to me, "Planting
tomatoes and lettuce may be the gateway to everything,
but not everyone can write books and stories and poems
as well as you do. Please don't waste your time with
manual work!" I have not wasted my time. Planting a
seed, washing a dish, cutting the grass are as eternal, as
beautiful, as writing a poem! I do not understand how a
poem can be better than a peppermint plant. Planting a
seed gives me as much pleasure as writing a poem. For
me, a head of lettuce or a peppermint plant has as much
everlasting effect in time and space as a poem.

When I helped found the University of Advanced
Buddhist Studies in 1964, I made a grave error. The
students, who included young monks and nuns, studied

47

only books, scriptures, and ideas. At the end, they had gathered nothing more than a handful of knowledge and their diploma. In the past, when novices were accepted into a monastery, they would be taken immediately into the garden to learn weeding, watering, and planting in full awareness. The first book they read was the collection of *gathas*, or practice poems, by Master Doc The, the book that included the poems for buttoning your jacket, washing your hands, crossing a stream, carrying water, finding your slippers in the morning, and other practical things so they could practice awareness all day long. Only later would they begin to study sutras and participate in group discussions and private interviews with the master, and even then, the scholarly studies would always go hand in hand with the practical ones.

If I were to help found another university, I would model it on old monasteries. It would be a community where all the students would eat, sleep, work and live everyday life in the sunlight awareness, perhaps like the Ark Community (L'Arche) in France or the Shanti Niketan or Phuong Boi communities. I am sure that in all the world's religions, meditation and study centres resemble one another. These are good models for universities as well.

Extract from The Sun My Heart *by Thich Nhat Hanh, reprinted courtesy of Parallax Press.*

Cultivating Living Intelligence

UNLEARNING THE CONCEPTS THAT CREATE THE ILLUSION OF SEPARATION

Vandana Shiva

"Seeing what is false is the Way of Intelligence."
– J. Krishnamurthy

We are members of one interconnected Earth Family of sovereign, autonomous, self-organised, interdependent, intelligent beings. We are one humanity on one planet. We are interconnected through the flow of breath and water, energy and food, and all life is rooted in community, communion, and communication. Indigenous cultures and wisdom traditions knew this, and emerging ecological sciences now recognise this; that the Earth is alive and gives us life. But we have been taught to forget our interconnectedness and oneness; we now see ourselves as separate from and superior to the natural world, so that we can deny that people and planet matter. Our dominant systems of science and technology are based in a militaristic, mechanistic paradigm and our economic model is founded in greed, consumerism, and the foolhardy illusion of limitless growth on a planet with ecological limits. This worldview has given rise to an existential crisis and multiple emergencies, such that, now, humans are a threatened species too.

These multiple emergencies are not separate – as with all things, they are interconnected, they have the same roots. Their solutions are also interconnected and can be found in the cultivation of 'Living Intelligence', enabling us to make peace with Mother Earth and co-exist harmoniously with all our relatives, both human and other-than-human. And we do this by re-imagining education. The word 'education' is derived from the Latin words *educare* which means 'to bring up' and *educere*, 'to bring forth'. We have the potential to bring forth our common humanity and our membership of one Earth Family as the foundation for all education. Cultivating this potential is now a survival imperative. We need an education that decolonises our minds, hands, and hearts by unlearning ways of thinking and acting based in separation and the illusion of superiority that have created the myriad emergencies we face.

Colonialism is based on myths of separation and superiority, and the conquest and control of the Earth and her diverse cultures, and a mechanistic paradigm was adopted to justify the exploitation of Nature. Education is still immersed in this mechanistic world view which denies and displaces Nature's creativity and intelligence. A fragmented, atomistic view of life was imposed on complex, interconnected living organisms and ecosystems, whereby individual life forms were assumed to be evolving in isolation and in competition with all others for scarce and shrinking resources, a perspective which is blind to the fact that the Earth and all beings are alive, cooperative, and abundant. It is a paradigm which asserts that Nature is dead and the Earth has no rights; that there are no ecological limits and no limits to extraction from Nature. It is an ecological apartheid and it is at the root of the myriad emergencies we face. Ecological apartheid allows for the emergence of anthropocentrism, the illusion that humans are superior to all other species and are masters of the Earth. We now have a collective amnesia of our heritage as one Earth Family.

As the great teacher, Krishnamurthy, once observed, "Seeing what is false is the way of Intelligence," and it is my assertion that Regenerative Education must be able to see what is false; we need an education that sees through the illusion of separation and the lies of dominion. Education for life has to cultivate our real intelligence, which co-creates with the intelligence of other beings and the Earth. Then, Nature becomes our teacher and through Nature we learn how to cultivate living intelligence, ecological intelligence, cooperative intelligence, compassionate intelligence, emotional intelligence. Trees and plants become our teachers, bees and butterflies become our teachers, mycorrhizal fungi become our teachers.

True knowledge of and immersion in the reality of an interconnected world of fellow beings is a wonderfully co-creative and radical act. It is a creativity that has been suppressed, subjugated and rendered invisible over the last few hundred years of colonialism and the dominion of mechanistic reductionism. But the Living Earth paradigm has re-emerged thanks initially to the Gaia Theory and those visionary scientists and thinkers who could see 'what is false'. Sciences of living systems are now rapidly evolving, and the epistemology of care and co-creation that has been at the heart of all Indigenous cultures is being recognised, even at this 'eleventh hour'. The knowledge systems of ancient cultures did not create an artificial division between humans and Nature, between mind and body. All Indigenous cultures talk of 'being Earth' of 'speaking Earth' and they possessed an unlimited imagination of how to co-create with Earth. Imagination without the boundaries of separation is at the heart of Living Intelligence.

When we are conscious of being members of a living Earth Family our knowledge shifts from the fragmentation and destruction of Nature's integrated living systems into resources for consumption, to an exquisite awareness of Nature's intelligence, of her living technologies and economies of abundance. As we unlearn the colonial concepts of separation and dominion, we learn how to open our hearts to natural wisdom. Our consciousness then shifts from a mechanistic monoculture of the mind to an abundant biodiversity of mind, based in interconnectedness, diversity and multiplicity; a non-violent epistemology of relationship, partnership, care and compassion. These are the qualities of Living Intelligence and of co-creating harmoniously with other life-forms in an interconnected world. When we are conscious that the Earth is living and that plants, animals and microbes are intelligent sentient beings, we learn to listen to them, and we are open to learn from them. When we break out of the anthropocentric and colonial construction of 'exceptionalism' we will see the intelligence of all life.

Living Intelligence is our compass to freedom from external control and manipulation.

Living Intelligence is at the heart of learning at The Earth University/Bija Vidyapeeth at Navdanya's Biodiversity Farm in Uttarakhand, India, inspired by Rabindranath Tagore, India's national poet and a Nobel Prize Laureate. Tagore started a learning centre in Shantiniketan in West Bengal, India, as a forest school, both to take inspiration from Nature and to create an Indian cultural renaissance. The school became a university in 1921, growing into one of India's most famous centres of learning. Today, just as in Tagore's time, we need to turn to Nature and the forest, to plants and biodiversity, for lessons in freedom. Forests are sources of water and the storehouses of biodiversity that can teach us the lessons of democracy — of leaving space for others while drawing sustenance from the common web of life. Tagore saw that unity with Nature is the highest stage of human evolution. It is this unity in diversity that is the basis of both ecological sustainability and democracy. Diversity without unity becomes the source of conflict and contest. Unity without diversity becomes the ground for external control. This is true of both Nature and culture. The forest is a unity in its diversity, and we are united with Nature through our relationship with the forest. In Tagore's writings, the forest was not just the source of knowledge and freedom; it was the source of beauty and joy, of art and aesthetics, of harmony and perfection. The forest teaches us union and compassion, it teaches us enoughness, as a principle of equity; how to enjoy the gifts of Nature without exploitation and accumulation. No species in a forest appropriates the share of another species. Every species sustains itself in cooperation with others. The end of consumerism and accumulation is the beginning of Living Intelligence. The conflict between greed and compassion, conquest and cooperation, violence and harmony that Tagore wrote about continues today. And it is the forest that can show us the way beyond this conflict.

Solutions are emerging because young people of today are feeling the call of the Earth – and eco-literacy is re-emerging in our minds, our hearts, our lives, and the way we produce and consume. These are the green shoots of an Earth Democracy that is taught through an education of Living Intelligence, and it allows us to protect the Rights of the Earth as well as those of human and other-than-human beings. Justice and sustainability are part of this interconnected process. It is possible, even now, to shift from the emergency-creating paradigms of patriarchy to Earth-based Living Intelligence that reduces our ecological footprint while expanding and deepening our hand-print, heart-print and head-print, connected as One Earth Family.

Vandana Shiva is an Indian scholar, environmental activist, food sovereignty advocate, ecofeminist and anti-globalisation author. Her most recent book is Reclaiming the Commons: Biodiversity, Traditional Knowledge and the Rights of Mother Earth.

THE TRUE MEANING OF EDUCATION

SCHOOLS AND UNIVERSITIES
CAN BECOME PART OF THE SOLUTION
RATHER THAN THE PROBLEM

Satish Kumar

*"Education should be so revolutionised as to
answer the wants of the poorest villager, instead
of answering those of an imperial exploiter."*
– Mahatma Gandhi

In the recent past, New York governor, Andrew Cuomo,
Bill Gates of Microsoft and former Google CEO, Eric
Schmidt, have been promoting the idea of transforming
face-to-face learning to a system of education rooted
in internet technology and operated by remote control.
Thereby, integrating digital technology fully and
permanently into the educational process, and by doing
so getting away from the need for personal relationships
and intimate interactions between students and teachers.
Cuomo, Gates and Schmidt come from a school of
thought which subscribes to the theory that, 'Technology
is the solution, what is your problem?' Unfortunately,
these highly 'educated' people do not seem to know the

meaning of 'education'. The word is derived from Latin 'educare'. It means to bring forth or lead out or draw out what is potentially already there.

Every human comes into this world with his or her own unique potential. The work of a true teacher is to observe and spot that special quality in a child and help to nurture it and enhance it with care, attention, and empathy. Thus, the beautiful idea of education is to maintain human diversity, cultural diversity, and diversity of talents through decentralised, democratic, human scale and personalised systems of schooling. A good school is a community of learners where education is not pre-determined by remote authorities, rather it is a journey of exploration where students, teachers and parents are working together to discover right ways to relate to the world and to find meaningful means of living in the world.

The idea of digital learning through remote control and pre-determined curricula moves away from the rich and holistic philosophy of education. Digital teaching looks at children as if they were empty vessels in need of being filled with external information. The quality of information or knowledge given to the child remotely and digitally is determined centrally by people who have a vested interest in a particular outcome. And that outcome is largely to turn humans into instruments to run the money machine and to increase the profitability of big corporations. Such centralised and impersonalised systems of digital education will destroy diversity and impose uniformity, destroy community culture, and impose corporate culture, destroy multiple cultures, and impose monoculture.

When teachers teach remotely, they tend to think as if the children have no body, no hands, and no heart. They have only a head. The information taught digitally is almost entirely of an intellectual nature. Thus, digitally educated children are less than half-educated. Eating half-baked bread gives you indigestion; similarly, life of a half-educated person lacks coherence and integrity. A proper education should include the education of the head, education of the heart and education of the hands.

Technology is seductive and a double-edged sword. It can be a useful tool to connect, as the pandemic has taught us through our use of Zoom, but it can also be a brutal weapon of control. If technology is the servant and if it is used with wisdom to enhance human relationships, without polluting the environment and without wasting natural resources then technology can be good. But if technology becomes the master, and human creativity and ecological integrity are sacrificed at the altar of technology then technology becomes a curse.

A computer cannot teach kindness. Only in a real learning community can children learn how to be kind, how to be compassionate and how to be respectful. In a school community, children learn together, play together, eat together, and laugh together. It is through these shared human activities that children gain a deep appreciation of life. Education is more than the acquisition of information and facts; education is a living experience. Sitting in front of a computer for hours is no way to learn social skills, and so placing the future of our children in the hands of a few digital giants like Google, Microsoft and Amazon and putting them in charge of educational systems is a recipe for digital dictatorship and opens the doors to disaster. If democratic societies are opposed to military dictatorship, then why should they embrace corporate dictatorship? Through smart technologies these giant corporations will be able to trace and exploit every activity of children and later, when they are adults, through data manipulation and control. Who wants to embrace such a 'dystopia'?

Placing the future of our children in the hands of a few digital giants like Google, Microsoft and Amazon and putting them in charge of educational systems is a recipe for digital dictatorship and opens the doors to disaster

Rather than investing in top down, artificial, sedative, and virtual technology, democratic societies should be investing in people. We should be investing in more teachers in smaller schools, with smaller class sizes and bottom up, imaginative, benign, and appropriate technology. We have already experienced the way algorithms, artificial intelligence, biotechnology, nanotechnology, and other forms of so-called smart technologies have been used to control, manipulate and undermine democratic values. Instead, we should be embracing the Green New Deal and not what Naomi Klein rightly condemns as the Screen New Deal.

We need the greening of education rather than the screening of education. Our children need to learn not only about Nature but from Nature. They need to learn from forests and farming, from permaculture and agriculture, from agro-ecology and organic gardening, from marine life and wildlife. Such knowledge and skills cannot be learned by looking at computer screens. A computer is a box. It teaches you to think within the box. If you want to think outside of the box, you need to go out into your community, and out into the natural world. Technology has a place in education but let us keep it in its place and not allow technology to dominate our lives and the lives of our children. Technology is a good servant but a bad master.

It seems the misunderstanding of the word 'education' can also be applied to the word 'economy'. Some time ago I was invited by the London School of Economics (LSE), to speak to their students on the subject, 'An Ecological Worldview'. It was my pleasure and honour to accept their invitation. Before my talk I was offered a warm welcome and hospitality. Over tea and cake, I asked my hosts, "Do you offer any courses for the study of An Ecological Worldview to your students?"

One of the professors replied, "We study Environmental Policy with Economics and also Climate Change and Economics, but no course on An Ecological Worldview as such."

I said, "Environment and ecology are not the same and climate change is a consequence of harmful economic growth, whereas the study of ecology offers knowledge, understanding and experience of the entire ecosystem and how the diverse forms of life relate to each other."

The professor replied, "That is too broad a concept. Our courses are much more specialised."

After this brief discussion we went to the Hong Kong Theatre where more than 100 students and lecturers had gathered in that delightful hall to hear me speak. I decided to build on the discussion I had just had, and this is the gist of what I said.

"The London School of Economics was a radical and pioneering university in the Age of Economy, but now we are entering the Age of Ecology. Therefore, LSE needs to respond to the needs of our time. LSE should become LSEE, The London School of Ecology and Economics. Let me explain. I am sure that you are all fully aware that ecology and economy are like identical twins: both words have Greek roots. 'Eco' or 'oikos' in Greek means 'home or household'; 'logy' or 'logos' means 'knowledge'; and 'nomy' or 'nomos' means 'management'. So, ecology is the knowledge of the household and economy is the management of the household.

"For Greek philosophers, home or household is a very inclusive term. A home is where we have our bedroom, living room, kitchen and bathroom. But a home is much more than that. A nation is also a home and ultimately the entire planet Earth is our home. And all the species of the Earth are interrelated, as one family. Amazing animals, fabulous forests, majestic mountains, awesome oceans and of course imaginative and creative human beings are all members of this Earth home, this planetary household.

"At LSE you teach economics. This means that you are teaching how to manage the Earth home. But you don't teach ecology. You are teaching your students to manage something without teaching them what it is that they are going to manage! LSE has taught thousands upon thousands of young leaders from around the world the techniques and methods of economic management. The world economy is in their hands. And sadly, it is not in good order. Actually, it is in a mess! But why should we be surprised to see that the world economy is in a mess? It is in a mess because the managers of the world economy don't know what they are managing!

"This is not only a problem at LSE. All the universities in every country around the world teach economics without teaching ecology. So, it is a problem of our entire educational system. The problem of teaching economics is even deeper. The curriculum and syllabus of courses in economics have very little to do with the study of Earth home management. Much emphasis is devoted to the management of money, and students have become little more than cogs in the money machine. While the various course titles may include the word 'economics', in practice there is very little reference to the management of the Earth home. Somehow, we have forgotten the true meaning of economy. Our focus has shifted from the management of the Earth home to the management of money and finance in the interest of a particular group of people rather than in the interest of all members of the Earth household.

"If we were to meditate on the original and actual meaning of the word, we would soon realise that the economy is a subsidiary of ecology. Without ecology there is no economy. Yet at LSE, as in other universities around the world, economy is taught as if there is no connection between economy and ecology.

"Nature, which is another name for ecology, is considered as just a resource for the economy which in effect means a resource for the maximisation of profit through ever increasing production and consumption. Thus, Nature has been reduced to a mere resource. Similarly, people are reduced to a resource for the economy. We call this 'human resources.'

And so, endless production, consumption, and the pursuit of profit, in the name of economic growth, progress and development have become the most cherished goals of the modern economy. Nature as well as people have become a means to an end; simply instruments to increase the profitability of businesses and corporations. According to an ecological worldview, production, and consumption as well as money and profit should be a means to an end, but the end goal should be the wellbeing of people and the integrity of the planet Earth. If production and consumption, money and economic growth damage Nature and exploit people, then such economic activities must be stopped at once.

"An industrial economy is a linear economy. We take from Nature, use it and then throw it away, with the consequence that it ends up in landfills, in rivers and oceans and in the atmosphere. We need to replace this linear economy with a cyclical economy. All goods and products must be recycled and returned to Nature, without waste. This is cyclical economy. In the economy of Nature there is no pollution. If we were to be guided by the wisdom of an ecological consciousness, we would work very hard not to produce any pollution. Pollution is a violation of the purity and beauty of our Earth home. With ecological sensitivity we would know that if we pollute the air we have to breathe, if we pollute the water we have to drink and if we pollute the soil from which we grow our food – then ultimately, we pollute ourselves.

"Without an ecological worldview, we humans consider ourselves separate from Nature and even superior to and above Nature. We value Nature only in terms of her usefulness to humans. Due to this anthropocentric attitude towards Nature, we have been on a centuries-long mission to conquer and subjugate Nature. This arrogant attitude is the root cause of the present ecological crisis. With an ecological worldview we change this attitude. We recognise the unity of human beings with all other living beings. We recognise the intrinsic value of all life, human as well as other-than-human life. As we uphold human rights, we also uphold the rights of Nature.

"The integration of ecology and economy is the urgent imperative of the 21st century. This is why I am urging LSE – and all universities – to embrace an ecological worldview. But more than that, I am asking LSE specifically to nail your colours to the mast by changing your name to the London School of Ecology and Economics, LSEE. In doing this, you will be making a statement that hereafter, all teachings at this university will be underpinned by the understanding of the Earth household and a proper management of it, where the interest of humans and those of Nature will be in total harmony. If universities wish to be part of the solution, then they can no longer ignore ecology.

Satish Kumar is the author of many books, most recently, Elegant Simplicity, *and is Editor Emeritus of Resurgence & Ecologist and Founder of The Small School and Schumacher College.*

The Age of Enchantment

ACTIVELY IMAGINING
EDUCATION SYSTEMS OF THE FUTURE

Stephan Harding

"Education is not preparation for life;
education is life itself."
– John Dewey

It is by now absolutely clear that our current educational paradigm is contributing to the destruction of biodiversity and to the associated deeply perilous shift in our global climate. Thankfully, more and more educators are now realising that this is because our educational processes promote the illusion that we humans are separate from Nature - that we are her masters and controllers. Mainstream education teaches us that Nature is no more than a dead machine without soul, intelligence or purpose and that we can exploit her 'resources' without let or hindrance to further the growth of our economic interests.

During my own university training in Zoology and Ecology in the 1970s and 80s, I was taught that selfishness and competition are the sole driving forces in the evolution of life and that these impulses are also at the core of our human natures. I well remember the depression and disgust I felt when forced to adopt the 'selfish gene' view of life so deftly promoted by neo-Darwinism as the deepest truth about the biological world. I strongly felt that selfishness alone could not have produced the marvellous diversity of life on our planet, but there were no persuasive counter-arguments available to me at the time, so I also felt alone. During my A-level Zoology course and later during my undergraduate degree in the same subject, we students spent most of our time indoors listening to lectures, dissecting dead animals and reading and writing in stuffy libraries with occasional field trips thrown in to count and measure animals caught, mostly dead, in our traps and nets.

It is now absolutely clear that our current educational paradigm is contributing to the destruction of biodiversity and to the associated deeply perilous shift in our global climate

All of this made me feel deeply unhappy, and it was only when I became a founding faculty member at Schumacher College that my feelings and intuitions about the severe limitations of my biological education were backed up by persuasive scientific arguments about the importance of cooperation and creativity in Nature, from great scientists who taught at the College, such as James Lovelock, Lynn Margulis and Brian Goodwin. At the college we were given the opportunity to integrate these insights with similar understandings from other cultures, other times, and other places, giving us a deeply holistic and very enlivening sense of Nature's ways and processes.

How might the education we have pioneered at Schumacher College over the last thirty years affect the world of the future? To explore this, I invite you to journey with me into a fantasy, or as psychologist C.G. Jung would say, into *active imagination*: we don't know who we'll meet, but whoever they are, let's allow them to reveal themselves to us, as in some sense real.

Let's imagine then, that by around the year 2421 the insights and educational practices developed at Schumacher College and in many other similar places around the world have matured and blossomed into deeply ecological modes of education sensitive to local cultures and ecologies. Gaia is in good shape again – she's regulating things very well now with her rich and wonderful mixture of human cultures happily embedded within her great wild biomes and regions.

Look! A young woman from those times is coming out of the future to meet us.

Oebphe: "Hello. I am Oebphe Ebepho, born in September 2392. I'm a 29-year-old graduate in holistic science. I studied at Cambridge University, England. I'm living on a wonderfully restored planet that has recovered well from the crisis our ancestors created, that almost pushed Gaia into a permanent hot state. Great wildernesses are everywhere now, full of astonishing biodiversity. Ecologically sustainable human cultures thrive like never before."

Me: "Greetings Oebphe. We are educators from four centuries before your time working to avert the global crisis that threatens us so dangerously. We would be so pleased if you would recount some highlights from your education. We need to implement your educational processes now in the 21st century so that you'll be able to enjoy the wondrous planet of your times, so full of life and wonder. We need to cultivate styles of education that will stop the planet lurching through a series of irreversible tipping points into a hot and far less habitable, far harsher world. Can you help us?"

Oebphe: "Of course. When we were little children, the focus was very much on myths and stories concerning the divinity and mystery of the cosmos, its grandeur and its wondrous origins. We were taught to pay close attention to our dreams and were encouraged to develop our own uniquely individual imaginal worlds out in the woods and gardens where our all our education took place. We spent no time at all in classrooms – all our education took place in the heart of Nature where we had roofed shelters with open sides for wet weather. Everyone was taught in small groups no larger than fifteen children. We did gardening, cooking and cleaning and learned to take care of each other and the Nature around us. This was our learning journey from about four to around eleven years of age."

Me: "Were you encouraged to analyse and think during those early years?"

Oebphe: "Not at all. That came much later, once each of us had grounded ourselves in our own well-developed, highly individual personal mythology of Nature and of our life within her. During those early years, we learned how to speak with plants, how to converse with birds, how to listen carefully to what the land was telling us about herself. We spent a lot of time tracking animals, making camps in the wilds, growing food in our gardens, telling good stories, connecting with the night sky, drawing and painting, sowing clothes and repairing things. When I was ten, I so clearly remember hearing the life stories of figures such as Goethe and Humboldt, whom we all loved so dearly and purely. We learned how to relate to rainbows and plants, to the whole of Nature, like they did. It was an awakening for us all, and it was a simple life, a profound education in what it is to be a young human, just eleven years old, alive in this marvellous world, so deep and rich in meaning, in friendship, connection and grandeur."

Me: "Did that give you a feeling of wholeness?"

Oebphe: "Oh yes. It was very integrating. We were deeply immersed in our sensory and intuitive relations with Nature. Our feelings were central too, and these were perhaps the most important, for they were full of love and wonder for the immense value of the physical world. We slowly became better at accessing these perceptions at any time, allowing us much later as young adults to react with great wisdom and compassion in any situation. I well remember making my first vow at that time to be in the service of our planet with all her beings and to the vast, intelligent universe, for the rest of my life."

Me: "That's remarkable."

Oebphe: "No it isn't. This is our normal. Everyone does it."

Me: "Did you work with numbers?"

Oebphe: "When we were twelve our teachers began to help us cultivate of our abilities for analysis and rational thought which had deliberately been put on the back burner until then. Our forays into analytical thinking took place out in the woods, fields, forests, mountains, in whichever biome we were in the world, in all weathers, under all conditions, so that we were always in the great presence of Nature, as much deeply immersed in her mythological dimensions as we were in these new and powerful ways of thinking."

Me: "What analytical tools did you learn?"

Oebphe: "We learned some beautiful mathematics which in your day you called 'complexity theory'. We were deeply struck by how simple equations (which we experienced as expressions of living relationships amongst the 'parts' of Nature) could produce beautiful patterns at the level of the whole that were so obviously the same patterns that Nature herself produces. We delighted in our ability to think clearly about these issues and easily integrated this new delight into our unshakable love of Nature as a creative living force, always for her benefit and in her service."

Me: "How long did this phase of your education last?"

Oebphe: "Until we were about fifteen, when we entered into our first rite of passage. Our teachers curated and mentored each of us personally through a five day-long vision quest in the depths of wild Nature, which is so wonderfully abundant everywhere nowadays. After the vision quest, we came back together in the woods. We were encouraged to reflect on any images that had come to us during our vision quests. We were amazed to find that most of us had seen a symbol with a fourfold structure of some sort, arranged symmetrically into what we learned was a mandala representing the deep Nature of reality and of human consciousness. We explored how such mandalas have helped people down the ages to cultivate the fullness of their humanity. Some cultures expressed this as Air, Water, Fire and Earth, others as North, South, East and West, others as Thinking, Feeling, Sensing and Intuiting. We realised the objective quality of the four-fold mandala and that it emerges out of a deeper form of non-dual consciousness which is that of the universe herself.

Me: "Did you feel a new sense of love and wonder as you reflected on your vision quest experiences together?"

Oebphe: "Very much so. They became deeper."

Me: "At what age did you begin to learn about the ecological crisis of our times and what had brought it about?"

Oebphe: "That was when we were about sixteen years of age. By that time, we had all done our first solo vision quest in Nature and had completed the first phase of discovering the full scope of our humanity and of our human potential. We were so deeply experientially grounded in the Soul of Nature, in the *anima mundi*, that we were ready to hear how you moderns, our ancestors, became so destructive of Gaia, our mother planet, whom we love so much.

"It was then that our teachers helped us to plumb deep into the depths of the human shadow and to understand how to engage in fruitful relationships with these darker aspects of ourselves. We went into Galileo, Descartes, the Scientific Revolution and all, so profoundly that we could feel the tragic pulse of those times in our bones."

Me: "They couldn't imagine that their new science would lead to the massive destruction of Nature."

Oebphe: "They might well have been horrified. How we wished we could have warned them of the great shadows they were unwittingly unleashing."

Me: "Did you and your friends at seventeen or so have a sense of where your lives were taking you, of how you would be of service Gaia and all her beings?"

Oebphe: "We all felt something very strongly – a calling. Some went this way, others that. Some became holistic medicine people, whilst others became engineers developing ever more efficient forms of generating energy sustainably. Some went into growing food ecologically in their local communities. There are a whole host of professions that are needed by the Gaian community, so we became a mycelial net of young people in the great web of Gaia with friendly professions available to us all."

Me: "And you chose to study biology at Cambridge?"

Oebphe: "Yes. It is wonderfully holistic there now, with the most modern learning from all sorts of fields united with the beauties of the mediaeval vision of life, with a world full of soul and mythology – with the *anima mundi*. At Cambridge and all our universities around the world we continue our journeys into the living heart of Gaia, no matter what we study."

Me: "And now that your education is over, what do you do?"

Oebphe: "My education isn't over..."

Me: "But you left Cambridge about a decade ago."

Oebphe: "Cultivating my connections to my immediate ecology, to the wild bush country, supporting local food growers and local enterprises, supporting my neighbours – all these things are vital for my ongoing learning – it can't develop without them. This kind of learning, which is so lovely and so delicious to experience, goes even deeper. I keep on learning about the wholeness of Nature in every aspect of my own life, through all my ups and downs, through thick and thin, through joy and pain."

Me: "And now?"

Oebphe: "Now I too am an educator. I take young people through the stages of learning I've been telling you about, although of course anyone can begin at any age. There's a worldwide network on the internet that's constantly exchanging people, ideas and experiences so we can cultivate all this more and more effectively. It feels so good to be rooted to my local ecology whilst being connected to the global community via this mindful use of technology."

Me: "Thank you Oebphe. Meeting you has been a great education. We have learned richly."

Oebphe: "And I loved meeting you, dear one. Our connection across time helps hasten the recovery of our planet's majesty and beauty which so enchants and enlivens us in our times. It's great to be alive in this moment of ours in the 25th century, which we know as the Age of Enchantment. Thank you for what your generation did to save all this for us in your difficult times, just in the nick of time, during your age of Re-Enchantment."

Me: "It was you. You inspired us."

Oebphe: "Goodness! I hear drums and guitars away over there by the great sacred Baobab tree. How lovely!

It's the dry season festival! It's beginning. I must go. The night is coming. All my friends will be there around the campfire. The owls are already hooting from the topmost branches. I will find you in your forest... Look for me there."

See now, how she's vanishing into the future, back into the beauty and richness of her days. Thank you, Oebphe Ebepho, from us all.

 *Stephan Harding is the Deep Ecology Research Fellow at Schumacher College. For almost two decades he was Head of the MSc in Holistic Science at the college. He has a doctorate in behavioural science from the University of Oxford, and is the author of* Animate Earth: Science, Intuition and Gaia, *and* Gaia Alchemy.

Profound Concern, Fierce Hope

TEN IDEAS TO IMPLEMENT
REGENERATIVE EDUCATION

Stephen Sterling

*"The volume of education has increased and
continues to increase, yet so do pollution,
exhaustion of resources, and the dangers of ecological
catastrophe. If still more education is to save us,
it would have to be education of a different kind:
an education that takes us into the depth of things."*
– E. F. Schumacher (1974)

The fate of the planet and of humanity hangs in the balance. Yet there is an astonishing disconnect between the pressing signs of global change and crisis, and the relatively closed world of education. Many people – particularly the young who are rightly concerned about their future – and international agencies such as UNESCO are recognising that education has been, and remains, part of the problem. It is time for radical change so that education can rapidly become a central part of the burgeoning global movement to reclaim, restore and regenerate human and natural systems towards the wellbeing of all and a safe future. There is increasing recognition that the *transformation of education* is needed if it is to play a full role in assuring the future. Hence UNESCO's current International Commission on the 'Futures of Education' emphasises the need to move towards a 'new ecological understanding' that 'integrates our ways of relating to Earth' and which 'requires an urgent rethinking of education'.

In this light, I put forward below ten key ideas which frame the argument for the rethinking and redesign of education from micro to macro scale, which I believe could help create a breakthrough transformation in educational policy and practice, enabling education to be transformative in its impact and fit for the future.

1. Seeing the world as it really is
The starting point is to acknowledge that all educational endeavour takes place within, is framed by, and reflects its cultural milieu. Therefore, problems of purpose and orientation in education need first to be seen within the larger context of the belief systems and assumptions that tend to define our culture.

Many writers over the years have argued that the systemic crises that characterise our age arise from the way the Westernised mind perceives, thinks, and values - that external dysfunctions are essentially a manifestation of a shared internal dysfunction in worldview. For years, our narrow 'system of concern' has been manifested at individual and societal levels in individualism, egocentrism, and anthropocentrism, whilst our sphere of influence and impact in the material world has been endlessly expansive. Now, with the Earth 'on fire', and signs of socio-economic and ecological breakdown becoming increasingly evident, the customary narrowness of our attention is no longer tenable. We are being forced to look both inwards; 'do our values, beliefs and accustomed ways of doing still hold?' and outwards; 'what is the larger context here?' – and to look at the relationship between the two.

All the warning signs suggest we urgently need to nurture an extended, inclusive, caring and holistic view of the world, whilst reducing and localising our physical and ecological impact, in order to be able to halt the damage and generate resilient systems into the future. *This requires a profound reversal of worldview and human impact: put very simply, a transition from a state of 'Small Worldview/Big Impact' towards a condition of 'Big Worldview/Small Impact'.*

2. Reframing learning

An emerging view is gaining ground that collapse of some sort is increasingly likely in future decades which may be ecological, economic, technological, social, or some combination of these. Some economists are predicting that biophysical limits will inevitably usher in a post-growth world characterised by re-localisation, profound hazards, and discontinuities for both human and natural systems.

In response, there's a new transformative and life-affirming zeitgeist arising, not least, driven by young people, and it's based on the twin drivers of profound concern and fierce hope. This heightened awareness is giving rise to some urgent re-thinking, realignment and refocussing in key areas of human activity. Importantly, this shift can be seen as a learning process – a social learning revolution, no less – which has three key aspects:

- a conscious and intentional *un-learning* of habits and ideas that got us to the point of systemic multiple crises
- a *re-learning* and reclaiming of approaches that are more sustainable and life-affirming
- *new learning* that has the potential to sustain human and natural systems wellbeing into the future.

Ironically, the education sector, which purports to prepare people for the future, has not, by and large, been at the forefront of this transformative zeitgeist, which is more apparent in the world of entrepreneurship and business.

3. Educational 'response-ability'
At a deep level, the prevailing educational paradigm reflects and is informed by the wider social paradigm or worldview. So, if the latter is maladaptive to contemporary global conditions, it follows that the prevailing educational paradigm will echo this orientation. The key here is what I call the *'response-ability'* of education, that is, the ability or otherwise of education systems to respond fully and effectively to the profound challenges facing us. It is poor response-ability in education that many students are concerned about and are drawing attention to.

Rather than 'education for sustainable development' (ESD) it is very likely we need 'education for sustainable contraction'. In the imminent post-growth age and faced with the immediate needs to mitigate and adapt to severe climate change, *the fundamental need in education is to be 'response-able', purposefully supporting the change from unbridled economic growth towards building resilience and flourishing within planetary limits, particularly at local and community level.*

4. Towards third-order learning

Education cannot be an effective agency of the social transformation needed unless the education community is itself transformed. The scope of this challenge is better understood if we make two sets of distinctions. The first set concerns *arenas of learning*: and distinguishes between structured curricular learning and institutional or organisational learning. Curricular learning is of course what all schools and universities support: that is, programmes for students; and it is this arena that institutions typically address when they wish to increase their response to sustainability issues.

The organisational learning arena concerns the social learning that policymakers, professional staff, and practitioners may themselves experience formally or informally. Sufficient attention here is critical to progress in the curricular arena, and also directly affects the possibility of whole institutional change including leadership and governance, partnerships, strategy, campus management and operations, research and curricula, and community engagement. In particular, universities, see themselves primarily as research and teaching organisations, not as learning organisations.

The second set of distinctions concerns *levels* of learning. This draws on Gregory Bateson's work who distinguished between three depths of learning and change. First-order learning refers to doing 'more of the same', that is, change within boundaries, and without examining or changing the assumptions or values that inform what we are doing or thinking. Second-order learning or meta-learning refers to a significant change in thinking or doing as a result of examining assumptions and values. Meta-learning is about recognising and understanding how our external practice is effected by our subjective world – beliefs, assumptions, and values.

From this distinction, it is possible to see that most learning promoted in formal education in schools and higher education is of the first order kind, being content-led and information-based. The pedagogy is transmissive, with little heed given to 'heart' and 'hand', the affective and the practical. Beyond these two learning levels, Bateson also distinguished a third order, which refers to paradigmatic change, that is, deep learning which is transformative.

Rapidly growing awareness of global crises is now precipitating second and for some, third order, social learning across society, but educational institutions still tend to be behind the curve in this realignment. Education must develop the 'response-ability' to embrace this third-order change.

5. Surveying the foundations
If we can get a better handle on how the prevailing educational paradigm reflects the wider cultural paradigm

or worldview, we can help facilitate second order – even third order – learning within educational systems.

Our cultural inheritance from the scientific revolution has been a worldview characterised by materialism, objectivism, reductionism, and dualism, underlain by a mechanistic metaphor. Our primary sense of reality is the physical world, which we try to understand by fragmentation, by looking at parts rather than wholes.

When institutions are faced with the question of how to respond to the challenges of sustainability, the normal route taken is to effect piecemeal rather than systemic change in provision and practice. This might, for example, extend to localised change in curricula or campus management. Whilst this is of value, I argue that sustainability requires deep attention to education itself, and as a whole — its interlaced paradigms, policies, purposes, and practices. Otherwise, it is rather like building an eco-house on old foundations that are no longer fit for purpose.

The foundations of the prevalent education paradigm exert a kind of hidden influence over purpose, policy, and provision and associated educational discourse. But if we see education and learners in mechanistic terms, it will always result – and has resulted – in the kinds of education systems that have prevailed in recent decades. So, how do we achieve the 'escape velocity' that is necessary to release education from the trap of outmoded and damaging ways of seeing, knowing and doing, towards modes that are caring, holistic, and regenerative and wise in practice? First, *we must loosen and challenge the economistic straitjacket that has been placed on policy and practice.*

6. Rethinking and reimagining

In the past 20-plus years, an already narrow educational paradigm has been compounded by an overtly instrumental view of education, informed by the rise of neo-liberalism in economics, politics, and wider society and by the perceived demands of a globalised economy, and the powerful rise of the 'EdTech' (educational technology) industry.

Marketisation and a 'global testing culture' have led to competition, homogenisation, and standardisation nationally and internationally. So educational discourse and practice has tended to work within these controlling parameters. This shift towards centralised control has squeezed out older liberal, humanistic, learner-centred and progressive concepts and models of education. This squeeze makes it harder for education policy and practice to address social and ecological wellbeing.

The narrowing influence of neo-liberal thinking on the purpose, conception and practice of education needs to yield to an expanded paradigm. But this argument is not new – E.F. Schumacher emphasised the need for education to remake and reimagine itself some 45 years ago, towards an education 'that takes us to the depth of things.' *A deep learning response within educational thinking, policymaking and practice is required, based upon an emerging relational or ecological worldview, already manifesting in diverse civil society movements.*

7. Catalysing 'positive dis-illusion'

The first step to transformative learning and change – on an individual, community, institutional, or societal level – is a necessary crumbling of key assumptions in the face of evidence that suggests they are increasingly untenable. These specious assumptions include that:

- we are essentially separate from the environment and Nature
- ecological systems are a subsystem of the economy
- economic activity must be paramount and indefinite economic growth is desirable and possible
- we will 'conquer' Nature
- Nature equals a stock of resources for human use
- the future is stable and assured
- science and technological innovation will solve all problems
- the next generation will enjoy a higher quality of life

Mounting evidence of systemic crises, with the volatile climate the most obvious sign, is catalysing 'dis-illusion' involving increasingly widespread questioning of these long-held beliefs, values and assumptions, and this process has now been accelerated by the onset of the coronavirus pandemic. For some, this process of dis-illusion is painful. It is reactive learning, where their worldview is uncomfortably challenged. But for increasing numbers of people, this an experience of positive dis-illusion, because it allows them to embrace exciting and new ideas of regeneration, innovation, and redesign that offer hope for the future.

This may be called *anticipative learning* – *which is reflexive, and critically self-aware, and is directed at wise, corrective action in the light of evidence. For increasing numbers, it involves more than a questioning of assumptions, but more deeply, the emergence of an alternative worldview which may be termed holistic, ecological, and systemic.*

8. Adopting a 'participative consciousness' worldview

Current crises – particularly gross inequity, conflict, environmental destruction and now a pandemic – are fundamentally systemic. So, a systemic reality necessitates a systemic (i.e., relational) or ecological worldview. The linking of human activity with the dangerous consequences of the heating of the Earth's systems is forcing learning amongst swathes of the public and decision makers, a late awakening towards a kind of systemic awareness that many environmentalists have been advocating for years.

Hence, the dawning of what is termed 'participative consciousness'; the awareness that everything really does relate to everything else and that all actions have consequences – from miniscule to massive, and from short term to long term. Far from being detached and unaffected observers, we are – unavoidably – participants inside a greater whole; we are not on the Earth, but in the Earth, inextricable actors in the Earth's systems and flows, constantly affecting and being affected by the whole natural/human dynamic relationship.

This worldview is ecological, life affirming, and regenerative and it's playing out in myriad social movements across the world. It is also inspiring great educational work, albeit often within the corners and margins of the mainstream – which now needs to come on board.

Adopting participative consciousness brings a shift of focus from relationships largely based on separation, control, manipulation, individualism, and excessive competition towards those based on participation, appreciation, an ethical sensibility, self-organisation, equity, justice, sufficiency, and community.

9. Embracing a higher purpose for education

Part of re-purposing education for our times involves reclaiming and reasserting some of its earlier focus on inner work, reflected in liberal and progressive traditions, and, with a foot on these foundations, moving towards modes that are less materialist, exploitative, short term and individualist than currently prevail.

We are not starting from scratch. For a long time, there has been a counter educational current reflected in such practices as progressive education, learner-centred, explorative and experiential education, emergent learning outcomes, liberal arts education, inter- and trans-disciplinarity, participative pedagogies, action research and co-inquiry, whole institutional change, community-based service learning, and more latterly, transformative, transgressive and transpersonal education.

These expressions of a more ecological and humanistic educational paradigm indicate ways in which the instrumental, neo-liberal control of educational policy and practice can be challenged, and a necessary degree of reclamation and reorientation can be achieved. Expression of holistic, relational education is always possible at some level, even in unpropitious circumstances, and by its practice the system change we seek becomes manifested.

10. Regenerative education is already happening

We have a short timeframe to ensure a liveable future for ourselves and other Earth species. This raises deep questions for virtually all areas of human activity but especially challenges education whose very *raison d'être* is preparation for life, and which is exclusively designed to affect values, dispositions, understanding, and competencies. The key is re-purposing, renewal, and regeneration, and developing a 'culture of critical commitment' in educational thinking and practice. There is evidence that it is happening.

Until recently, I was co-chair of the UNESCO–Japan ESD Prize International Jury. In that role, I've read close to 500 outstanding submissions from formal and non-formal projects across the globe. They reflect a heartening level of energy, commitment, inventiveness, courage and determination to empower people to make a positive difference to their locales and spheres of influence. This is a kind of authentic, rather than commodified, education that is already achieving a difference through many projects and initiatives. Small and independent institutions such as Schumacher College in Devon, UK, are playing an exemplary role in this pioneering work, and the important role of such centres of radical learning has been out of all proportion to their size.

In these practices, education as a vehicle of social reproduction and maintenance is superseded by a vision of continuous re-creation or co-evolution where leading edge education on the one hand, and progressive movements in society on the other, are engaged in a relationship of mutual social and individual transformation – with the common good and wellbeing of the human and the more-than-human world in mind. Fundamentally, this is about rediscovering our humanity and creating a new story on this still beautiful planet.

> Young people are fervently waiting for us educators to catch up with them, to empower them and to help them make the future

With the signs of the 'Great Unravelling' increasingly evident, societies need to manifest and enact 'the Great Turning' as coined by Joanna Macy – which has been alternatively labelled 'the Great Work' (Thomas Berry), or 'the Great Transition' (Paul Raskin). The work is underway but needs acceleration. Urgency necessitates building agency and capacity – and that is one thing on which education can claim a degree of monopoly.

There's a fork in the road, and education has to make a choice. Young people are fervently waiting for us educators to catch up with them, to empower them and to help them make the future. *By offering a holistic way of seeing and being, a regenerative, ecological, life-affirming paradigm can help shape new and energising pathways for a more hopeful, and secure future.*

 *Dr. Stephen Sterling is Emeritus Professor of Sustainability Education at the Sustainable Earth Institute, University of Plymouth, UK. He is a Distinguished Fellow of the Schumacher Institute, a Senior Fellow of the International Association of Universities (IAU) and a Fellow of the Environmental Association for Universities and Colleges (EAUC). His research interest for many years has been in ecological thinking, systemic change, and transformative learning at individual and institutional scales to help meet the challenge of accelerating the educational response to the sustainability agenda and crisis. www.sustainableeducation.co.uk*

Ecoliteracy and the Dance of Cooperation

We Need to Teach Our Children, Our Students and Our Political and Corporate Leaders the Fundamental Facts of Life

Fritjof Capra

*"Forget not that the earth delights to feel your
bare feet and the winds long to play with your hair."*
– Kahlil Gibran

It is evident today that concern for the environment is no longer one of many issues; it is the context of everything — of our lives, our businesses, our politics. The great challenge of our time is to build and nurture sustainable communities, designed in such a manner that their ways of life, businesses, economies, technologies and social institutions respect, honour, and cooperate with Nature's inherent ability to sustain life.

The first step in this endeavour, naturally, must be to understand how Nature sustains life. It turns out that this involves a new ecological understanding of life. Indeed, such a new understanding of life has emerged in science over the last few decades. At the forefront of contemporary science, the universe is no longer seen as a machine composed of elementary building blocks. We have discovered that the material world, ultimately, is a network of inseparable patterns of relationships; that the planet as a whole is a living, self-regulating system. The view of the human body as a machine and of the mind as a separate entity is being replaced by one that sees not only the brain, but also the immune system, the bodily tissues, and even each cell as a living, cognitive system. Evolution is no longer seen as a competitive struggle for existence, but rather as a cooperative dance in which creativity and the constant emergence of novelty are the driving forces. And with the new emphasis on complexity, networks, and patterns of organisation, a new science of qualities is slowly emerging.

I call this new science 'the systems view of life' because it is based on systemic thinking, or 'systems thinking', by which I mean thinking in terms of relationships, patterns, and context. Thinking in terms of relationships is crucial for ecology. That word — derived from the Greek *oikos* or 'household' — is the science of the relationships among various members of the Earth Household. I should also mention that systemic thinking is not limited to science. Many Indigenous cultures embody profound ecological awareness and think of Nature in terms of relationships and patterns.

In modern science, systems thinking emerged in the 1920s from a series of interdisciplinary dialogues among biologists, psychologists, and ecologists. In all these fields, scientists realised that a living system — an organism, ecosystem, or social system — is an integrated whole whose properties cannot be reduced to those of smaller parts. The systemic properties are properties of the whole, which none of its parts have. Thus, systems thinking involves a shift of perspective from the parts to the whole. The early systems thinkers expressed this insight in the now well-known phrase, 'The whole is more than the sum of its parts.'

During the 1970s and 1980s, systems thinking was raised to a new level with the development of complexity theory, technically known as 'nonlinear dynamics.' It is a new mathematical language that allows scientists for the first time to handle the enormous complexity of living systems mathematically. Chaos theory and fractal geometry are important branches of complexity theory. The new nonlinear mathematics is a mathematics of patterns, of relationships. Strange attractors and fractals are examples of such patterns. They are visual representations of the system's complex dynamics.

One of the most important insights of the systemic understanding of life is the recognition that networks are the basic pattern of organisation of all living systems. Ecosystems are understood in terms of food webs (i.e., networks of organisms); organisms are networks of cells, organs, and organ systems; and cells are networks of molecules. The network is a pattern that is common to all life. Wherever we see life, we see networks. Indeed, at the very heart of the change of paradigms from the mechanistic to the systemic view of life we find a fundamental change of metaphors: from seeing the world as a machine to understanding it as a network.

Closer examination of these living networks has shown that their key characteristic is that they are *self-generating*. In a cell, for example, all the biological structures — the proteins, enzymes, the DNA, the cell membrane, etc. — are continually produced, repaired, and regenerated by the cellular network. Similarly, at the level of a multicellular organism, the bodily cells are continually regenerated and recycled by the organism's metabolic network. Living networks continually create, or recreate, themselves by transforming or replacing their components. In this way they undergo continual structural changes while preserving their web-like patterns of organisation. This coexistence of stability and change is indeed one of the key characteristics of life. All life is inherently regenerative.

Now let us return to the concept of ecological sustainability. To understand how Nature sustains life, we need to move from biology to ecology, because sustained life is a property of an ecosystem rather than a single organism or species. Over billions of years of evolution, the Earth's ecosystems have evolved certain principles of organisation to sustain the web of life. Knowledge of these principles of organisation, or principles of ecology is what I call 'ecological literacy', or ecoliteracy for short.

In the coming decades the survival of humanity will depend on our ecological literacy — our ability to understand the basic principles of ecology and to live accordingly. This means that ecoliteracy must become a critical skill for politicians, business leaders, and professionals in all spheres, and should be the most important part of education at all levels — from primary and secondary schools to colleges, universities, and the continuing education and training of professionals.

Ecoliteracy must become a critical skill for politicians, business leaders, and professionals in all spheres, and should be the most important part of education at all levels

We need to teach our children, our students, and our political and corporate leaders the fundamental facts of life — for example, that one species' waste is another species' food; that matter cycles continually through the web of life; that the energy driving the ecological cycles flows from the sun; that diversity assures resilience; that life, from its beginning more than three billion years ago, did not take over the planet by combat but by partnerships and networking.

All these principles of ecology are closely interrelated. They are just different aspects of a single fundamental pattern of organisation that has enabled Nature to sustain life for billions of years. In a nutshell: Nature sustains life by creating and nurturing communities. No individual organism can exist in isolation. Animals depend on the photosynthesis of plants for their energy needs; plants depend on the carbon dioxide produced by animals, as well as on the nitrogen fixed by bacteria and fungi at their roots; and together plants, animals, and microorganisms regulate the entire biosphere and maintain the conditions conducive to life. Sustainability, then, is not an individual property but a property of an entire web of relationships. It always involves a whole community. This is the profound lesson we need to learn from Nature. The way to sustain life is to build and nurture communities.

As soon as the first cells appeared on Earth, they formed tightly interlinked communities, known as bacterial colonies and for billions of years, Nature has maintained such communities at all levels of life. Natural selection favours those communities in which individuals act for the benefit of the community as a whole. Indeed, the patterns of organisation that ecosystems have evolved sustain a vast network of relationships with countless forms of cooperation, partnerships, and symbiotic arrangements. They are all forms of behaviour for the common good.

In the human realm, such behaviour for the common good is known as ethical behaviour. It includes the goals of social justice and peace, which are values of human consciousness and culture that are not relevant in non-human ecosystems. Social justice and peace are sometimes called 'social sustainability'. For a human community to survive and flourish in the long run, it needs to be both ecologically and socially sustainable. Both ecological and social sustainability are realised in human behaviour for the common good.

This is an enterprise that transcends all our differences of ethnicity, culture, or class. The Earth is our common home and creating a sustainable world for our children and for future generations is our common task.

Fritjof Capra, Ph.D., is a physicist and systems theorist (www.fritjofcapra.net). He is Founder of The Center for Ecoliteracy (www. ecoliteracy.org) and author of The Tao of Physics *and co-author of* The Systems View of Life.

Reimagining Schools

WE ALL HAVE A 'GREAT DREAM' FOR OUR CHILDREN'S SCHOOLING. WHAT'S YOURS?

Anthony Seldon

"Don't be pushed by your problems.
Be led by your dreams."
– Ralph Waldo Emmerson

We are all far more influenced by our education then we might care to admit. So, it matters deeply what our young learn at school, how they learn it, and the quality of their relationships with fellow students and staff. Kindness and empathy can be encouraged and developed. A love for fellow human beings, for animals and for the natural environment can be nurtured and acquired. Equally, a narrow materialistic outlook and self-centred behaviour can also be generated if the right stimuli and experiences are not offered.

We reflect too little and rarely on our own experience of school, and when we are encouraged to do so, we often return to well-trodden memories which may or may not be true. So, when I suggest that you make a list of the best school experiences you had for developing your love for humanity and your harmony with Nature, I hope you will neither pass on to the next paragraph, or chapter, nor simply trot-out a list that is familiar to you. Instead, try to tap into memories that are fresh, because doing so will aid the purpose of this article, and help us to reimagine schools for the future, but which we need to manifest now.

> There is much more that is right about the education system than we often acknowledge. It is full of remarkable schools and teachers, and educational standards that are clearly improving, with the inner child nurtured more than was the case even 20 years ago

Imagine you can design a school from the ground up, where money is no object, and there is no given curriculum to deliver, with the freedom to select the kind of teachers and support staff who you think will best help the children learn, and the community from which they come; imagine a school where caring for others and the natural world is the norm. Then come up with a list of perhaps 10 ideas which you would like to see instituted, drawing on your own experiences. Then set about implementing them!

My own list is inspired by the 10-part acronym 'Great Dream' from Action for Happiness, the charity that I helped found ten years ago with Lord Richard Layard and Geoff Mulgan. We looked for two words which would spell out the 10 steps or actions which we believed would lead to happier people and communities. Before I give you my own list of 10 ideas, let me first say that I believe there is much more that is right about the education system than we often acknowledge. It is full of remarkable schools and teachers, and educational standards that are clearly improving, with the inner child nurtured more than was the case even 20 years ago.

Let's now see how 'Great Dream' might be applied to designing more humane and green schools.

GIVING: The first letter is G, which stands for 'giving'. It is an insight widely shared amongst positive psychologists, that happy people are givers, whereas unhappy people are takers or withholders. Supposing this simple idea lay at the heart of every school? When we think of giving, most think of money, which is not unimportant, but there are so many other ways in which every one of us could give more. Take energy, for example. Those who give energy freely to others, letting energy flow freely through them, are happier and more in touch with other human beings. Those who block energy, or let it turn inward, lose opportunities for connecting with others. Ultimately, they become depressed and stale. The light within them can no longer shine out in their self-absorption. 'Energy is eternal delight' was a phrase of poet WB Yeats which I remember recording in my special book at school: he was right, but only when we let the energy circulate. We can also give others our time, our attention and love. Schools which ground themselves deeply on the experience of giving, letting the young taste the fulfilment it imparts, will be preparing them well the world they will face.

RELATING: R stands for relating. All schools should all be grounded in good relationships. A badge or a school motto declaring love is not sufficient – love has to be lived from the ground up. The Head needs to be humane, approachable, and kind to all. They should set the tone for the entire school in their every action because everyone else models their behaviour upon him or her. All teachers need to live the message. Children who are encouraged and treated with respect, and who are given clear boundaries, are much more likely to grow up well-rounded and balanced than children who have been disparaged and abused at school. The more difficult the child's home, and the more starved of love and boundaries, the more important school becomes.

EXERCISE: Our third letter, 'E' stands for exercise, which should lie at the heart of every school. Good schools should educate the body as well as the mind. This is easier if schools have wide open spaces, but even compact urban schools can have daily yoga classes for the children, physically stretching and strengthening their bodies. Strong bodies support strong minds. Learning about the body needs an earlier focus too, where all children are taught how to relax deeply and learn about the importance of good nutrition, proper hydration, and what they should put into their bodies, and avoid putting into their bodies. A child who is taught how the body works will have a proper respect for it for the rest of their life. I'm so glad that daily star jumps, stretching and bending were part of my own schooling from the age of six in South London. But later, health and safety concerns, or extra maths lessons (I forget which) took their place.

APPRECIATION: Our fourth idea can and must be taught at school. The development of a half-empty, or half-full mindset, an optimistic or a negative outlook, is established long before children leave school. In the last few years, with the positive psychology work of Professor Martin Seligmann based at the University of Pennsylvania, some schools are being encouraged to use 'gratitude diaries', in which the students record what they are appreciative of. Another practice is thinking about 'Three Blessing' before going to sleep. In this exercise, children are encouraged to reflect on three events during the day for which they are grateful. Without such an exercise, habitually negative and anxious thoughts can invade the mind on the liminal borders of sleep, but intentionally thinking of happy events can replace them, giving a greater depth of sleep and slowly rewiring the brain away from ingrained negative patterns of thought.

TRYING OUT: Schools should all be encouraging the spirit of adventure, suggested by our fifth letter 'T', standing for 'trying out'. Adventure and challenge should be hardwired into every school. This spirit can apply in formal lessons, where pupils and staff should be encouraged to experiment and occasionally rip up the rulebooks, but it applies particularly outside the classroom, where all students from a young age should be encouraged to have adventures in Nature and spend decent chunks of time in open air. Where possible, overnight camping trips should be organised, with hikes and explorations in different scenery. Pushing the young beyond their comfort zones and allowing them to achieve success in overcoming self-doubt and negativity. This a great gift for young people.

When I was head at Wellington College in Berkshire, one of the very best activities with the Lower Sixth was to take them out at 10 pm to walk 25 miles along The Ridgeway, the ancient Icknield Way through the heart of England. There were always huge groans and protests, and a few predictable last minute dropouts, but the vast majority went, hated it, then loved it and never stopped talking about it – then or since. Timbertop, which is part of Geelong Grammar School in Australia, takes all its young teenagers away for an entire term into the outback, with no mobile phones. That life-changing experience is beyond the pocket of most schools, but it is not beyond them to capture something of that same spirit and excitement.

Ashley Primary to the south-west of London, led by the inspirational head Richard Dunne, pioneered many of Prince Charles' ideas of Harmony: the children grew some of their own food and nurtured animals, while learning about the principles of harmony that underlie the universe. The remarkable 'Country Trust' offers experiences for city children from disadvantaged backgrounds to visit farms, where they learn about healthy cooking, living in Nature, and seeing farms in action. The experience is utterly transformative, as is Michael and Clare Morpurgo's 'Farms for City Children', founded in 1976. Tim Smit's Eden Project, which is sprouting branches across the UK and world, provides different opportunities for children to understand Nature. They all have so much to give the young that they will not learn in the standard academic curriculum.

The second word, 'Dream' suggests five further actions which could be embedded into every school to help turn out young people who are more humane and in tune with Nature.

DIRECTION: comes from the first letter, 'D'. A sense of direction helps the young to develop orderly lives in place of disorder and chaos that so many experience. By giving them a sense of harmony in their lives, it will help them to live more manageable and happier existences. Having a sense of direction and order is much more likely to lead to a happier and productive life than one which is out of control.

RESILIENCE: The letter 'R' in this instance refers to the resilience of the wider surroundings in which the students are active. Their environment needs to nurture them and be inspiring. School buildings ideally should be constructed using resilient, sustainable natural materials from the local area, with local craftspeople and builders responsible for building them, and demonstrating their building techniques to the students. Edwin Lutyens, the celebrated British architect of the early 20th century believed strongly in using local materials and local craftspeople for schools. He wanted his buildings to be organically rooted in the environments in which they were situated. He would have hated the anonymous monstrosities of new schools imposed by central diktat with a uniform style and using dead and alien materials.

Emotional resilience can also be instilled in the young, whose lives will inevitably be affected by hardships, reversals of fortune and sadness. The more inner strength that we can impart to young people, the greater they will to be able to sail through turbulent times without sinking. The work of Angela Duckworth of the University of Pennsylvania has much wisdom and practical advice to give schools about the development of grit and resilience among the young.

EMPATHY: The 'E' of 'Dream' is empathy which lies at the heart of every successful and complete human being. Empathy is easier to develop when schools are not overly large. Children need to know and be known. Primary school should not exceed two-form entry, which means that no primary school would be larger than 500 pupils. Senior schools should not exceed four-form entry, and no school should be larger than 1000 students. It is impossible for a Head to know more than 1000 students, and their families and circumstances. I know from my own experience of 20 years as a Head how important this first-hand knowledge is; one person in every school has to be able to hold the whole community in the palm of their hand, and they cannot do that if they do not know everyone. Within schools, students should belong to 'Houses' which have a meaningful purpose, with excellent pastoral carers who know each child deeply, and what will bring the best in them.

AWARENESS: The penultimate letter, 'A', stands for awareness. From the earliest age, children learn how to do, not how to be. Their minds are cluttered with desires and baggage, and they find it harder and harder as they grow older to know who they are. Children as young as three, four and five should always have a period of calm at the start of each day, while older children can have periods of stillness punctuating the day, at the start of each lesson, in assembly or before sports practice. Helping the young to develop their own inner stillness, to have a place that they can go to which is safe and still when they feel frightened, or lonely, or afraid or nervous, is a wonderful gift to give each child.

MEANING: Finally, and most appropriately, the final letter 'M' stands for meaning, something schools are not very good at providing. Some years ago, I came across a Manchester primary school which ran the social and emotional learning programme called 'Zippy's Friends'. The children were taken by a teacher to a graveyard. It struck me as macabre when I first heard about it, but the more I learnt, the more intrigued I was, until I started inviting the Head who ran the programme to speak at wellbeing conferences for schoolteachers. A sense that life has no meaning is ubiquitous amongst the young, and indeed those of all ages. To help offer a sense of meaning from the earliest age, young people need to reflect on what gifts they have, what they love, and how they might give back to life. Visiting a cemetery and reading on the gravestones about the lives of those who have passed away gives a snapshot of how each person's life and can be inspirational.

Yes, school really matters! It is quite likely that you will produce a much better list of ten actions for more humane and environmentally-aware schools than mine. What matters now though is implementing the ideas that you have. It is you who will remake the world.

Sir Anthony Seldon is one of Britain's leading contemporary historians, educationalists, commentators and political authors. He was Vice-Chancellor of The University of Buckingham from 2015 to 2020 and is author or editor of over 45 books on contemporary history. He was the co-founder and first director of the Institute for Contemporary British History, is co-founder of Action for Happiness and is Deputy Chair of The Times Education Commission. He is a director of the Royal Shakespeare Company and President of IPEN, (International Positive Education Network).

Educating for a New Economy

THE SILENT REVOLUTION THAT IS REVERSING CENTURIES OF DESKILLING

Helena Norberg-Hodge with Henry Coleman

"Tell me and I may forget; show me and I may not remember; involve me and I will understand."
– First Nations Proverb

The Dean of Stanford University, Ellwood Cubberley wrote in 1934, "Our schools are, in a sense, factories, in which the raw materials – children – are to be shaped and fashioned into products. The specifications for manufacturing come from the demands of 20th-century civilisation, and it is the business of the school to build its pupils according to the specifications laid down." Thankfully, there has been a marked shift in many countries around the world away from the industrial, patriarchal values enshrined in Cubberley's factory model of schooling. In various spheres of life – from schooling to farming to home – there's been an embrace of more spiritual, humane, and ecological values. We are seeing diversity, creativity and indigeneity celebrated. We are witnessing a growing desire among all kinds of people to weave deeper relationships with

Nature and with one another. Even in mainstream school systems, these changing values are emerging, as inspired teachers pioneer curriculum reforms and more holistic approaches.

However, there is still an entrenched link between education and the economic system. And the economic system – despite the fact that cultural values have shifted – has continued in the direction of ever-larger-scale, globalised extraction, and exploitation of people and the natural world. It's vital that we understand this link and raise the call for a systemic shift in direction towards smaller-scale economic structures and place-based knowledge systems. Such structures are needed to respect the living diversity of ecosystems and of individual human beings and are essential to support genuine human and ecological wellbeing.

Despite societal demand for more ecological development, over the past three decades governments have ratified a series of 'free trade' treaties that has put the economy in the hands of increasingly large and powerful corporations. The 'big is better' ideology of these corporations (and of the policymakers they influence) has accelerated resource-extraction and destructive development across the entire planet. In step with this acceleration, western-style schooling has been exported to every corner of the globe. It encourages extreme individualism and competition, and trains people in the technological skills and worldview they need in order to be admitted to the global economic order.

This schooling stands in stark contrast to Indigenous ways of education, which tend to be intergenerational, collaborative, and experiential. This became clear to me after living in the traditional culture of Ladakh in the 1970s and 80s – before the region was subjected to westernised 'development' and outside economic control. In traditional Ladakh, children grew up surrounded by people of all ages, from young babies to great-grandparents – they were part of a whole chain of relationships. In this intergenerational context, I saw how they were naturally inclined to help one another – the eight-year-old teaching the five-year-old, who in turn lent a helping hand to the two-year-old.

Children learned from grandparents, family, and friends — often by *doing* — by imitating their elders. They learned and experienced connections, process, and change: the intricate web of fluctuating relationships in society and the natural world around them. From toddlers to grandparents, everyone took part in producing food, helping with irrigation, harvesting, and herding animals, as well as building houses gathering fuel, and a host of other tasks essential to survival. Such skills required manual dexterity and great physical strength, along with a keen, creative mind and intellect. It also required location-specific, ecological knowledge, that allowed them, as they grew older, to use resources in an effective and sustainable way. In short, education nurtured an intimate relationship with the living world.

This all began to change when western-style education came to Ladakh in the 1970s. In the modern schools, none of the cultural or ecologically adapted knowledge was provided. Children were instead trained to become specialists in a technological monoculture. School was a place to forget traditional skills and, worse, to look down on them. Children were fined for speaking their own language instead of English or Hindi. They were told by their teachers that their traditional culture was backward. Meanwhile, they learnt how to measure the angle that The Leaning Tower of Pisa makes with the ground and struggled with English translations of the *Iliad*.

To this day, school curricula are very similar – from Ladakh to London, from New York to Nairobi. Thanks again to the economic pressures that come from the global economic order, studies are increasingly focused on things like 'STEM', computer programming, chemical manipulation, public-relations, and business management: skills that grease the cogs of what can be described as a global techno-economic machine. But even more important than what is included is *what is left out*. The world over, modern education is educating people out of the skills they need to survive and thrive in their own cultures and regions, omitting any mention of generations of ecological knowledge and practical, land-based skills. At the same time, it fails to promote wellbeing and holistic thinking, and actively removes people from full participation in the real, living world.

Again, this has an economic imperative so basic that it is almost completely unconscious: If people are trained out of self-reliance and into dependence on a global consumer economy, corporate profits grow. In the modern education system, agricultural 'experts' don't learn how to nourish people – they learn about chemical-intensive techniques to mass-produce commodities for export. Engineering and architecture graduates don't learn how to build houses from the materials that are abundantly available in local bioregions – they learn how to build using a narrow-range of energy-intensive manufactured commodities. The result is artificial scarcity. People worldwide end up competing for a narrow range of globally traded commodities — from fossil fuels and rare earth metals to wheat, corn, cement, and plastic — while ignoring the relatively abundant resources of their own ecosystems.

Thus, in cities like Beirut and Dubai, air-conditioned skyscrapers with windows that don't open are replacing traditional forms of architecture that are designed for passive cooling. Across the world, housing developments are constructed of imported steel and plastic, while trees on site are razed and put through the woodchipper. Even as the use of terms like 'regenerative', 'biodiversity' and 'ecological' abound, the biological deserts of industrial monocultures are replacing diversified, localised, sustainable food systems. Such examples epitomise the ecological blindness of over-specialised knowledge for universal application. They also illustrate the systemic escalation of resource-use, pollution and waste that emanates from a globalised, industrialised economy. Through the education-industrial complex, global replaces local, urban replaces rural, monoculture replaces diversity, consumerism replaces culture, fossil fuels and technology replace people, competition replaces community. This system is also leading to social breakdown. As our communities fracture, and as competition for scarce jobs and resources rises, financial and psychological insecurity escalate. This globalising system is taking us away from Nature, away from each other, and away from ourselves.

As I mentioned at the outset however, the schooling system has thankfully not been able to quash the intuitive intelligence of countless people around the world, who are doing their best to steer education and the economy in a very different direction. Despite widespread westernisation, industrialisation, and urbanisation – or, actually, *in reaction to them* – there is a growing recognition of our innate spiritual connection to community and to Nature. Similarly, people are realising that learning essential skills like growing food are beneficial for the development of both body and mind.

> The schooling system has thankfully not been able to quash the intuitive intelligence of countless people around the world who are doing their best to steer education and the economy in a very different direction

A vital movement can be seen growing at the grassroots on every continent, even as it pushes against the grain of the dominant economic system. It involves a huge range of initiatives – from permaculture-based education to Steiner and Montessori, from community-schooling to wilderness-immersion experiences. Meditation, collaboration, art, music, and dance are respected and practised. Forest schools, edible school yards, classroom kitchens, and other more practical forms of education, abound. All such projects – broadly termed 'alternative' – offer more place-based, hands-on, experiential, ecological, collaborative educational experiences. They represent a small, relatively silent revolution, reversing centuries of de-skilling. Whether consciously or not, this alternative education movement is acting in tandem with transformative steps towards a new economy – towards more spiritual, Nature- and community-based, localised ways of life. It imagines a future in which human beings are much more than left-brains performing narrow, manipulative tasks in front of screens. By allowing human hands, hearts, and minds to act as a cohesive, creative whole, this grassroots movement is preparing people to contribute to communities that have over-sight and control over their own food systems, their own economies, and their own lives.

In the words of one First Nations Nishnaabeg scholar, referencing the Nishnaabewin knowledge system, which is now undergoing a process of revitalisation, "It does not prepare children for successful career paths in a hyper-capitalistic system. It is designed to create self-motivated, self-directed, community-minded, interdependent, brilliant, loving citizens, who at their core uphold our ideals around family, community and nationhood by valuing their intelligences, their diversity, their desires and gifts and their lived experiences."

In the light of learnings from Ladakh and from many other Indigenous cultures, an education system that ensures the future wellbeing of our planet and of its people will look more ancient than modern. That transition can start at multiple levels. For example:

- In early education, parents and community members can be actively included. Kindergartens can be brought together with old people's homes in symbiosis – perhaps in outdoor settings and in community gardens. In this way, the elderly can receive more of the connection they need to remain fit and healthy, while children are allowed to play more freely in Nature, under watchful, caring eyes.
- In higher education, more interdisciplinary modes of investigation can be funded, and the status of experiential knowledge can be elevated, vis-à-vis the current focus on information deduced through abstracted, reductionist datasets. Specialisation and corporate funding can be challenged and rolled back.

In saying that, genuine systemic transformation can only be achieved if many more of us participate in a rather different kind of education. We call it 'education-as-activism' or 'big-picture activism'. This is not about educating our children, but about educating ourselves to see the connections that are not obvious in our anonymous global structures. 'Education-as-activism' involves:

- Unlearning some basic assumptions that have been drummed into us through schooling and mainstream media. For example, questioning the ideas of 'progress' and 'development', and revising our understanding of human nature, wellbeing and knowledge.

- Becoming more literate in economics, by developing a basic understanding of the skewed subsidies, taxes and regulations that have abetted global corporate expansion.
- Examining how the large-scale, globalised economic system inevitably increases resource-use, while widening the gap between rich and poor and jeopardising personal and societal wellbeing.
- Learning about the profound ways that place-based, localised economies can heal our societal and environmental wounds. Seeing why smaller scale economic and political systems – based on transparent, reciprocal relations – are needed to respect biological and cultural diversity.
- Reawakening our deeper spiritual selves. Pausing to listen to our hearts and our embodied wisdom. Allowing ourselves to feel the joy of connection to the deeper dance of life, to let go of the need to master, to control, to label and 'know'.

This is about resisting the dominant ideology and actively disseminating a different vision. We need our analytical, 'left-brain' faculties to examine and articulate the flaws in global economic structures, and to go up against mainstream 'experts' and prevailing dogmas with robust argumentation. In doing so we need to be guided by a deeper, more intuitive knowing, an embodied wisdom that is leading us towards genuinely empathetic and joyous relations with ourselves, with others and with all of creation.

 *Helena Norberg-Hodge is a pioneer of the new economy movement, and a recipient of the 'Alternative Nobel' prize, the Arthur Morgan Award and the Goi Peace Prize. Author of the inspirational classic* Ancient Futures, *she is also producer of the award-winning documentary* The Economics of Happiness. *Helena is the founder and director of* Local Futures *and* The International Alliance for Localisation.

Henry Coleman helps to coordinate Local Futures *projects in Ladakh and Australia. He speaks at public events and works on various writing projects.*

WORLDLY EDUCATION

LIBERATING LEARNING FROM
THE CATEGORIES THAT CONTAIN IT

Charles Eisenstein

*"Education is a natural process carried out by the
child and is not acquired by listening to words
but by experiences in the environment."*
– Dr Maria Montessori

To align education with people and planet is not a
matter of merely changing its curriculum. Education's
basic categories – that of school, for example, and of
teacher, and student – themselves bear the patterning of
modern society as we have known it. Our organisation
of learning is counterpart to our organisation of the
rest of the world, human and otherwise. Today, as the
myths and meanings of modernity crumble, we have the
opportunity to liberate learning from the categories that
have long contained it.

Let us start with the very concept of education. The
word bears a different valence than 'learning,' doesn't it?
Fundamentally, learning is something you do; education
is something done to you. I learn. I am being educated.
While we might excavate its etymology for a different

possibility, in its normal sense education already foretells the institutionalisation of learning.

We call this institution of learning 'school.' The great countercultural intellectual Ivan Illich observed, "The very existence of obligatory schools divides society into two realms: some time spans and processes and treatments and professions are 'academic' or 'pedagogic,' and others are not. The power of school thus to divide social reality has no boundaries; education becomes unworldly and the world becomes uneducational."

As the term 'ivory tower' demonstrates, we well recognise the unworldliness of education today. Students speak of leaving school to enter 'the real world.' As well they might, after 16 years or 20 years of education in which everything they have done is but an exercise or a lesson whose worst possible consequence is a bad grade. By insulating children from real-world risks and responsibilities, it is no wonder that schools produce infantilised adults who act, choose, and consume oblivious to the consequences for the real world. The sequestering of children into schools and the containment of education in classrooms are part of the modern dissociation of human from Nature, and even from materiality. Knowledge is reduced to quantity, people into consumers, Nature into resources. Mind is separated from heart and from hand.

This unworldliness, even anti-worldliness, of education seemed a good thing in an age when civilisation's guiding myth was that we were destined to dominate, conquer, and rise above Nature. Now, as the paradise that conquest promised turns to hell, we realise that we cannot serve the world by transcending it.

We must question education as a category
separate from other social processes

Reworking the curriculum around social and ecological values amounts to putting new wine in an old skin. Changing the pedagogy is a step forward, yet it still risks preserving key categories and relationships that need to change. Pedagogy, after all, is the method of *teaching*. We must also look at the physical and social environment where teaching happens, and whether teaching is always necessary and appropriate for learning. We must also question education as a category separate from other social processes.

Let us broaden what we mean by education to include the processes by which one becomes a responsible, capable, mature adult. Responsible: responding to the gifts and needs of the world on every level, including those of community, place, and planet. Capable; possessing the skills that the needs of people and planet call for. Mature: to become a fully sovereign being connected to a purpose beyond the self. Within all these qualities, self-sovereignty is fundamental. Sovereignty is not the separate neoliberal individual's arrant disregard for the needs of the community and the world. True sovereignty is to fully own one's presence, role, and responsibility for the realm. That requires being embedded in community – in the broad sense of full, robust relationships to the rest of human and non-human life. The self is no longer separate and responds as the embodiment of those relationships.

One aspect, therefore, of education for an ecological and just world is that it restores and develops relatedness. Again, this is not a mere matter of curriculum – teaching about relationships, ecology, and so forth. It requires, in fact, quite the opposite of what teaching has been and where it has taken place. It requires restoring worldliness to education.

Certain lineages of alternative education point the way to such a future. Montessori education, for example, breaches the wall of separation between the classroom and the 'real world' with project-based learning. Crucially, when properly practiced these projects are not mere academic exercises, like making a diorama of the Pleistocene or writing a research paper, both of which expire upon grading and remain in the world, if they remain at all, collecting dust in the closet. The project must be real: starting a business, planting a garden, organising a theatre. It is not insulated as classroom exercises are from causes and consequences. It thus prepares the student for a life likewise responsible for causes and consequences.

Waldorf education offers another line of inspiration. It holds off on any kind of abstraction until at least the age of seven, focusing first on physical activities like cooking, puppets, singing, sewing, and making things. When academics are introduced, they begin in physicalised form: music, art, dance, recitation, and so on. Rather than compel students through force of authority and academic bribes and threats, Waldorf education holds that engaging with the beauty of the world is what inspires academic effort.

A third element of future education may be found in democratic free schools, pioneered by Summerhill in England and the Sudbury Valley School in the United States. These schools typically have no grades, no tests, and no curriculum. Staff are not allowed to teach unless invited to do so by the students. The school meeting, in which each student, even as young as five, has a vote, makes school rules, distributes budget, and hires new teachers. Here, students learn the skills of community rather than the habits of obeying and evading authority. They prepare for a life in which they create their own curriculum rather than marching through a life laid out by institutions.

This last example brings up a significant paradox. In a sense, conventional schooling is already worldly – it prefigures in its unworldliness the society that contains it. Yes, modern society is unworldly, so drunk with its spectacles and so lost in abstractions, particularly that which we call money, that it actively destroys the very world upon which it depends. The current educational system is a (perverse) success, preparing people to contribute to such a society.

A healed society is a worldly society, a society fully responsible to its members and the rest of life. To heal means to become whole, to return to integrity, to reintegrate. A worldly education in the true (not the perverse) sense is one that reintegrates school with society, reintegrates children into the life of the community, and reintegrates teaching with other professions. All these categories will become more fluid, as their lines of demarcation blur. The classroom will dissolve into the community and the ecosystem. Study will dissolve into apprenticeship. Education will dissolve into life.

Defining features of modern schooling (exams, grades, competition, papers, classrooms, disciplines, etc.) will not disappear, but will recede to a diminished role. Academic education will perhaps return to its original purpose – the training of scholars – while shedding its industrial-age aim of breaking children to the routines and tedium of the factory. The scholar dwells in a world of abstractions, and today so does the policymaker, the financier, and the technologists, all immersed in a sea of zeros and ones, pixels and images, dollars, and cents. Let us leave abstractions to the scholars and reconnect the rest of education, in both content and in form, to the body, the soil, the emotions, the water, the human being, and the life of the Earth. That is the re-worlding of education, a crucial part of civilisation's return to human and ecological reality.

Charles Eisenstein is a renowned speaker and author. His books include Sacred Economics, The More Beautiful World our Hearts Know is Possible *and* Climate: A New Story.

Spheres of Influence

THE CIRCULAR ECONOMIES OF EARTH SYSTEMS NEED TO BE AT THE HEART OF EDUCATION

Herbert Girardet

"Humankind does not experience itself as a part of Nature but as an outside force destined to dominate and conquer it. Humans even talk of a battle with Nature, forgetting that if we won the battle, we would find ourselves on the losing side."
– E. F. Schumacher, *Small Is Beautiful*

We have an existential problem: humanity has become an immensely powerful, planetary force and much of our education system is focused on the goal of building a prosperous future for ourselves, based on unrelenting economic growth, yet with current practices we are endangering the very future of humanity and the natural world. Few of us are empowered to engage with this reality. Now, as never before, we need an education and information system that provides plausible new perspectives on how we can live in peace with life on Earth. Faced with a planetary emergency, is ecological transformation, or even metamorphosis, of modern society still a possibility, and where do we start?

The space on Earth that contains life, including human life, is generally called the 'biosphere,' so named by Austrian geologist Eduard Suess in 1875. He wrote, "The plant, whose deep roots plunge into the soil to feed, and which at the same time rises into the air to breathe, is a good illustration of organic life, interacting between the upper sphere and the lithosphere. On the surface of continents, it is possible to single out an independent *biosphere*." This concept was revisited in 1926, when Russian/Ukranian biochemist Vladimir Vernadsky published his book, *Biosphere*, focussing on the interaction between planetary biology, chemistry, and geology. He said, "The biosphere is the only region of the Earth's crust where life is to be found. ... Without life, the face of the Earth would become as motionless and inert as the face of the moon." The term biosphere is now familiar to all as the place where photosynthesis reigns supreme, enabling the vast diversity of life to exist.

In 1979, James Lovelock, in his book, *Gaia: A New Look at Life on Earth*, further developed this perspective, arguing that, "life maintained the stability of the natural environment, and that this stability enabled life to continue to exist." His hypothesis stimulated the emergence of 'Earth Systems Science,' focussed on both the interlocking cycles of Nature, and the human interactions with it. Lovelock's Gaia Hypothesis has given rise to a new 'Earth consciousness' at a time when the relationship between humans and our home planet is becoming ever more precarious. I would submit that Earth consciousness needs to become the basis of a philosophy of education that can help us develop a holistic view of the evolving relationship between people and planet.

Earth consciousness is of critical importance at a time when human impacts on the biosphere have reached unprecedented proportions. A study recently published by Professor Ron Milo of the Weizmann Institute in Israel found that human activities have reduced the biomass of wild marine and terrestrial mammals by six times and the biomass of plant matter by half. Farmed poultry today makes up 70 per cent of all birds on the planet, with just 30 percent being wild. 60 percent of all mammals on Earth are livestock, mostly cattle and pigs, 36 percent are human and just four percent are wild animals. "I would hope this gives people a perspective on the very dominant role that humanity now plays on Earth," professor Milo told *The Guardian*.

All this has occurred very recently. For most of our existence we lived as hunter-gatherers, small in number, existing as part of a largely unaltered biosphere and utilising a very limited range of tools. Human dominance has been driven, above all else, by the evolution of technology. Fast forward into the 21st century, and we have superimposed a *technosphere* on Nature. A population of 7.7 billion people, equipped with a vast array of new technologies, makes unprecedented resource demands.

The term technosphere was first coined in 1968 by the Vancouver-based control engineer John Milsum. It comprises all the structures and processes that humans have imposed on the planet – from building technology, factory production, transport and communication systems, to whole cities, and on to waste dumps and many kinds of environmental pollution.

The impacts of the technosphere on the biosphere are also well illustrated by the weight of accumulation of inanimate matter that has occurred: to procure the materials required for our urban-industrial structures requires extraction of minerals as never before. A research team at Leicester University has recently quantified the actual physical weight of the technosphere, consisting of concrete, stone, tarmac, and metals, at an astonishing 30 trillion tons. This compares with just 546 billion tons of carbon in the world, of which 82 per cent is plants. One of the team, Professor Jan Zalasiewicz, says, "The technosphere may be geologically young, but it is evolving with furious speed, and it has already left a deep imprint on our planet."

Of course, much of this has been fuelled by the use of non-renewable fossil fuel deposits accumulated in the Earth's crust over hundreds of millions of years. Our current rate of combustion of oil, coal and gas has been quantified as around one million years' worth every year.

It is important for education to convey a clear understanding of the systemic difference between biosphere and technosphere:

- The biosphere, driven by solar energy and photosynthesis, is an essentially circular system, which is all about reproduction, organic growth, species interdependence, and regeneration. All wastes are recycled into new growth, assuring continuity of life.

- The technosphere, largely powered by fossil fuel combustion, is an essentially linear system. It is defined by resource extraction, mechanical production, chemical manipulation, and linear waste disposal, with pollutants accumulating in the biosphere, systemically undermining the continuity of life.

As Schumacher once said, "The system of Nature, of which man is a part, tends to be self-balancing, self-adjusting, self-cleansing. Not so with technology." In its current form, the technosphere clashes with the functional principles of the biosphere, as an organic, ecologically defined system. The biosphere is characterised by *negentropy* – sustained order – whereas in the technosphere *entropy* – emergent disorder – is writ large.

Above all else, it is the technology-derived lifestyle of the newly emerging *homo urbanus* that is largely unrestrained by knowledge about his environmental impacts. It is of the greatest urgency that this profound deficiency is addressed by new approaches to the education curriculum, including history teaching.

Faced with the ecological crisis that is upon us, we urgently need new historical perspectives. Before the industrial revolution the human economy was essentially circular, with steady food supplies as a central concern: organic wastes, in particular, were used as the basis for sustainable farming systems. Today, the vastly expanded human economy has become essentially linear: wastes are discharged into Nature as pollutants, to the detriment of future life.

Environmental externalities are largely unaccounted for, with dire consequences.

The good news is that some of the new tools made available to us by the technosphere enable us to glean new perspectives. Using satellite technology, we can now see the Earth from space, and it is an astonishing vista. At night, when darkness used to prevail, much of the Earth is now brightly illuminated by billions of lights from individual households, public buildings and vehicles. The lights of cities and the flares of oil and gas fields are turning night into artificial day. These images vividly illustrate the unprecedented human presence on Earth.

Then, during daylight hours we can see straight lines and right angles stretching across vast landscapes. Cities with their angular building blocks and multi-lane highways are much in evidence, mostly located along sea shores and rivers. Elsewhere we see vast fields utilised for large-scale mechanised agriculture, often sprawling across former forest landscapes. In some places there are feedlots with tens of thousands of cattle crammed together behind impenetrable fences. Elsewhere, green circles are clustered together in large numbers – the patterns of irrigated crops imposed on otherwise infertile landscapes. These are the ecological footprints of an urbanising world.

Cities in one part of the planet are linked to distant farmland, forests and mines to still their insatiable appetite for resources. To convey an understanding of these global 'teleconnections' is an important task for education at all levels, junior education and university tuition, as well as further education and life-long learning.

The total extent of the Earth's surface is 51 billion hectares, with 71 percent ocean, and 29 percent land surfaces. Agriculture uses half that land, with forests now covering 30 percent, about two thirds of the area at the end of the last ice age, though most of these are ecologically impoverished. And importantly there are few references to the fact that the planet's *living surfaces* extend to a vastly larger area than its *land surface*. Vernadsky estimated the active leaf surfaces of Russian forests to being between 22 and 38 times the land surfaces on which they grow. But walking in a tropical rainforest, and looking up at its multi-layered canopy, with innumerable epiphytes, it is obvious that its leaf surface may be far larger than the leaf area of trees in temperate forests. Precise estimates are still not available.

The Gaia Hypothesis states that the composition of the Earth's atmosphere is kept at a dynamically steady state by the presence of life, assuring its continuity. Large scale deforestation has been occurring in the tropics, and this interferes with this capacity. With rapid loss of tropical forest cover and marine vegetation now occurring, the Earth's bioactive surfaces are being continually reduced. This is particularly problematic at a time when biological carbon sequestration, countering CO_2 accumulation, is more important than ever, yet only half of our CO_2 emissions are currently being absorbed by photosynthesis. Most tragically, the Amazon Forest, a vital organ of planet earth, is becoming a net emitter of CO_2.

The degree to which deforestation has caused the loss of vast layers of living vegetation has barely been estimated. In fact, apart from supplying timber and new land for agriculture, it also opens access to mineral resources. The ever-growing demands of the technosphere, as it currently operates, undermine the very capacity of the biosphere to absorb our waste discharges, whilst also interfering with the Earth's water cycles, nutrient cycles, and carbon cycles. It is becoming increasingly apparent that much of economic growth across the world, based on depleting the integrity of the biosphere, has effectively become uneconomic growth: deforestation, resource depletion, pollution and climate change inevitably damage the relationship between people and planet. Prevailing economic theory and practices are clearly failing much of humanity. It is vital for our education system to convey this reality.

The systemic principles underpinning the industrial revolution, which gave rise to the technosphere, are now on trial: our industrially-powered economic system, which puts economy before ecology, largely ignores critical environmental externalities. And our legal systems have barely started to acknowledge the concept of 'ecocide', pioneered by environmental lawyer Polly Higgins, which highlights the ongoing damage inflicted on the biosphere.

Back to Vladimir Vernadsky: in collaboration with French theologian and philosopher Teilhard de Chardin, he also pioneered the concept of the *Noosphere*, described as the 'planetary sphere of reason'. "The noosphere represents the highest stage of biospheric development, its defining factor being the development of humankind's rational activities."

Some people argue that the development of cyberspace – the virtual space created by the internet – resembles or even represents this noosphere, which could instil a new Earth consciousness in humanity. Could the global information dissemination, enabled by the internet and global communication, be the key educational tool to help us get to grips with the fundamental mismatch between the biosphere and the technosphere?

It is certainly true that the internet enables unprecedented access to information, knowledge, and even sources of wisdom. So far so good. But as a powerful communication system it is being increasingly usurped by commercial interests, as a tool of a new global 'surveillance economy'. In the face of this, every possible effort should be made to build up the internet as a vital tool for disseminating knowledge of our home planet.

At a time of planetary emergency, education and the dissemination of information about life on Earth cannot just be focussed on young people. With little time left to prevent the Earth from overheating, rapid decisions have to be taken to overcome the systemic conflicts between biosphere and technosphere. All of us, including decisionmakers of all descriptions, have to undergo a crash course in Earth systems science. Concepts and practices for enabling the regeneration of the Earth's living, organic economy should be at the heart of education, focused single-mindedly on long-term wellbeing of people and planet. The need for us to reinvigorate the vital organs of the biosphere is closely linked to economic practices that enable us to lead less demanding, simpler lives.

All of us, including decisionmakers
of all descriptions, have to undergo a
crash course in Earth systems science

The growth of the eco-technical revolution, and the 'green new deal', now under way, may be a step in the right direction, but can it avoid dependence on metals such as lithium, cobalt, coltan and copper, for use in renewable energy devices and electric vehicles? Many of these minerals are mined in places previously covered in forest ecosystems.

It was the American biologist Barry Commoner, a founder of the modern environmental movement, who defined four laws of ecology in his pioneering book *The Closing Circle*, in 1971. They are as relevant as ever:

1. Everything is connected to everything else. There is one biosphere for all living organisms and what affects one affects all

2. Everything must go somewhere. There is no 'waste' in Nature and there is no 'away' to which it can be thrown

3. Nature knows best. The absence of a particular substance from Nature is often a sign that it is incompatible with the chemistry of life

4. Nothing comes from nothing. Exploitation of Nature always carries ecological costs and these costs are significant

These four laws have never been more relevant. They remind us that we must act according to principles rather than just convenient pragmatism. They inspire a coalition of young campaigners and seasoned, older voices who are coming together to develop concepts for a wholesome, regenerative relationship between us humans and our home planet. But now, some people even say we should pack our bags and try making a home on sterile, deep-frozen Mars. How mad is that?

Nature, as a vast, multifaceted, interactive living system, needs to be our teacher. Its economics are very different to current practices. Nature is all about give and take, in a dance of life powered by sunlight and wetted by rain. This circular system, in which all organic wastes are invariably recycled into new life, contrasts with the global takeaway systems we have installed: to take and take, but not to give. Until we adopt Nature's 'circular' ways, always giving back what we have borrowed, we will deplete her resources to a point where human life itself is in question.

All this points to the need to create an education system which is focussed on ways and means of restoring the Earth's living, organic economy. This should be seen as central to 21ˢᵗ century teaching and learning.

Professor Herbert Girardet is a cultural ecologist, working as an international consultant and author. He is a member of the Club of Rome, the World Future Council and the World Academy of Art and Science, and an honorary fellow of the Royal Institute of British Architects. He is trustee of Resurgence

and is a recipient of a UN Global 500 Award for outstanding environmental achievements. He has developed sustainability strategies for London and as 'Thinker in Residence' for Adelaide, South Australia. He has produced 50 TV documentaries on environmental topics for major broadcasters. His 13 books include Earthrise (1992), The Gaia Atlas of Cities (1992 and 1996); A Renewable World – Energy, Ecology, Equality (2009), *and* Creating Regenerative Cities (2014).
Website: www.herbertgirardet.com

THE SYNERGY OF ARTS AND SCIENCE

WE MUST CHANGE OUR WAY OF THINKING FOR EDUCATION THAT IS FUTURE-FIT

Donald Gray

"A change in the way of seeing,
means a change in what is seen."
– Henri Bortoft

We are facing a crisis on the Earth. A crisis caused by humanity that will require a solution from humanity, but it cannot be a solution based on the way we currently think. The way we currently think, and the way that science operates, has resulted in the destruction of our own life support system on this planet. Nine planetary boundaries have been identified by scientists themselves to delineate 'the safe operating space for humanity'. We have already transgressed four of these: biodiversity loss and extinctions, chemical pollution, climate change and land use change. Science and technology have allowed us to exploit the Earth's resources as never before, thus exposing us to dangers that science can now describe but did not anticipate. This has changed the face of the Earth to such a degree that we are now said to be living in a new geological era, the Anthropocene. Such dangers are so evident that the world-renowned theoretical physicist Stephen Hawking suggested that science was becoming a threat to our existence. But what are the new ways of thinking that we need to counterbalance and inform the dominant scientific mode, which has become so prominent in shaping our lives, and the Earth, today? It is perhaps difficult to imagine a new way of thinking, a different way of doing science and science education, but Henri Bortoft, a writer on physics and the philosophy of science, has argued that indeed there are other ways of thinking and doing science.

Perhaps it is best to start with a brief and necessarily simplified look at how science works. The modern scientific institution is built on the idea of separating out the object of study from its context, *the environ*, as if it were no longer connected neither with its surroundings nor with its observer. The object, so identified, is then divided into smaller and smaller parts of study, a process known as reductionism; crucially, any human quality of experience is removed from that process in order to consider only those features that can be measured quantitatively. This process inevitably treats humans as outside of Nature. Nature is something over there, while human beings are over here. There is no doubt that this reductive and quantitative approach in science has provided us with immense progress in medicine and other areas, which has improved the quality of life for many and the treatment of what were at one time fatal or debilitating diseases. Science and technology have also enabled us to make vast quantities of consumer goods and wealth, but it has resulted in the destruction of our natural environment and the loss of vast numbers of species which we also depend on to maintain the fragile web of life.

There is hope, however, in that our understanding of the way in which natural systems function has improved and new ways of approaching science are being developed. While the reductive approach in science is founded very much on a Newtonian mechanical model, we now know that this model does not work effectively for complex, open systems, the way in which living and natural planetary systems operate. This moves us away from the reductive and quantitative approach to one that must take account of non-linear feedback loops, greater uncertainty and more holistic systems thinking. Yet school and university science education is still heavily dominated by the one approach to science – that of the reductive and quantitative – and rarely addresses the limitations of such an approach.

To change our way of thinking
we need to change our way of educating

The recognition of complex systems has led to renewed thinking in science. For example, when issues are complex and stakes are high, such as environmental degradation and biodiversity loss, the scientists Silvio Funtowicz and Jerome Ravetz proposed the idea of post-normal science. This involves participation and input from an extended peer community of all those who have some legitimate knowledge and perspectives, which can include local and Indigenous knowledge, that can inform the decision-making process. This approach differs from policy being guided purely based on the dominant reductive science. Similarly, there has been an increasing interest in the idea of sustainability science, which recognises the importance of multiple interacting levels in natural systems and focuses on the dynamic interactions between Nature and society. While these can be viewed as positive moves in addressing some of our complex socio-environmental issues today, the scientists working in these areas have, nevertheless, been educated in a traditional scientific way, which is still dominant in our institutions. There is an argument that to change our way of thinking we need to change our way of educating.

I want to diverge briefly from science and contemporary science education to consider some developments in understanding how we learn. Recent research and thinking indicates that 'body-mind-environment' interactions play a fundamental role in cognition, a theory known as enactivism. Enactivist approaches to cognition emphasise the role of the dynamic coupling of body-mind-environment and as a result, movement and sensory engagement are important in this process, as well as the affordances made available from our environment. The role of the body was recognised over one hundred years ago by John Dewey who would use the hyphenated body-mind to indicate the essential and integrated nature of the two, but in enactivism we bring in the idea of action in and with the environment. The role of dynamic interaction with our environment, particularly the natural environment, has long been recognised by prominent educators over the centuries, such as Comenius, Montessori, Pestalozzi, Froebel and many more. Yet today, the pressures on children and young people from a school system that is primarily focused on economic and employability agendas, has resulted in largely static, classroom-based learning, heavily modelled on the idea that all learning takes place primarily in the head and is aimed at success in examination systems. Children and young people are being educated to be servants of the economic state and are losing an already fragile connection with and understanding of the natural world. A natural world, which we, and the economic structure of society, ultimately depends on. Without Nature, without ecology, there is no economy.

Returning now to science education I want to consider what a renewed science education might look like, if we want a science education that educates children and young people about humanity's position within the natural world. This science has to develop a recognition of the interconnectivity and fragility of our natural world and our place within it. A new science and a new way of thinking will have elements of the post-normal and sustainability science mentioned previously, but it must also result in a transformation of scientists, so that they feel themselves to be intricately connected to the web of phenomena that they are investigating. They, and we, are not separate from Nature and we need to be able to truly understand what the impact of decisions and actions based on science will be. We need to be able to develop a sensitivity to our place in the natural world so that actions we take that lead to the destruction of the natural world will be experienced as self-destruction. We need to cultivate an attitude which seeks to participate with Nature, rather than attempting to control it. We need to experience the incredible wholeness of which we are a part, which brings us to an approach to science that has had increasing interest in recent years. The science of Goethe.

Goethe is perhaps more widely recognised as an artistic, literary, figure but he also wanted to be recognised for his contribution to science and it is his methodology that has aroused particular interest. Goethe's way of science is said to be highly unusual because it seeks to draw together the intuitive awareness of art with the rigorous observation and thinking of science. Goethe's mode of understanding also sees the parts in light of the whole, fostering a way of science that dwells in Nature. Using his artistic sensitivities in his approach to science, Goethe developed a way of investigating phenomena, of investigating Nature, which enabled insights that had hitherto been thought not possible with human intellect. One of the fundamental principles of Goethe's method was sensory involvement in close observation of our environment, building on that experience with more systematic and rigorous studies. The synergy created through the fusion of arts and science, it is suggested, can bring about new insights, which, if coupled with appropriate sensory experiences in natural environments, can also develop a profound relationship with our natural world. The arts expose viewers to new ways of seeing, feeling, and thinking about Nature, and this can lead to greater awareness of and motivation to act on behalf of Nature. As the eco-artist, Andy

Goldsworthy, said, "All forms are to be found in Nature, and there are many qualities within any material. By exploring them I hope to understand the whole."

The importance of the arts is being increasingly recognised through hybrid forms of curricula seeking to bring forth greater dialogues amongst different ways of knowing and being in the world. One such example is STEAM education. STEAM seeks to integrate the STEM (Science, Technology, Engineering and Mathematics) with the Arts. It must be said, though, that the earlier drivers of STEAM education were premised on economic arguments and not on ecological ones. However, if we think of learning as the coordination of action within one's environment – like John Dewey encouraged us to do – STEAM education coupled with sensory, experiential education in outdoor natural environments could be a powerful new way to stimulate changes in the way we think and relate to the natural world.

While the term STEAM refers to science, technology, engineering, arts, and mathematics, I wish to use the letters of the acronym to reconfigure an approach to science and education, which encapsulates some of the fundamentals that may be able to change the way we approach education, and science education in particular, and perhaps, change our way of thinking. I provide a brief reconfiguration here, which has been elaborated elsewhere in relation to the activity of gardening and STEAM gardens.

S for SENSORIAL: Science education, in fact education as a whole should enable children and young people to use their senses as much as possible. Too often science is confined to measurement and detached observation. When in a garden, for example, focus on sensorial engagement in the first instance, rather than measurement and testing. Rather than just sifting soil to separate particle size, or measuring pH, or allowing sediment to separate in a jar of water, let children get their hands in to feel the different textures, let them smell the richness of organic soil, let them observe the different colours of different soils, let them taste the products of the Earth, let them hear the sounds of Nature. Focus on the senses and how they feel, rather than identification, classification, labelling and measuring. Let them enjoy the wonder of the living world. The 'facts' can come later, but first let them connect and wonder.

T for TIMELY: Here we need a change from the idea of time as *chronos* to the idea of time as *kairos*. Chronos is time as measurable and sequential, divided into seconds, minutes, hours and days, the times of the day and the year. It is scientific time. It is quantitative time. Kairos, however, is the recognition of the appropriateness of time, the right time for action. Kairos is qualitative. Schooling is ruled by chronos; it is determined by dates in the calendar, school term times, timetables and exam dates, all of which are mediated through a technology designed to speed things up. Kairos, in contrast, recognises that, just as there are the right times to do things in the garden, it is important to recognise the appropriateness of certain moments in the education of children. In the current environmental crisis, there has never been a time when it has been more appropriate to take the necessary action and to educate young people about why these actions are necessary.

E for Enactive: Recognising the importance of the interconnectedness of body, mind, and environment, we need to develop pedagogies and approaches that make the most of the affordances available in the environment, particularly the natural environment, and allow children and young people to move and be active. This idea should form one of the foundations of our educational systems.

A for AESTHETIC: Just as arts enable us to see differently, we need to cultivate the aesthetic sensibilities in children and young people. We need to develop an understanding of the intrinsic worth and beauty of Nature, even in its wildest forms.

M for MATTER: We now are more aware that matter matters. Things, objects, materials are not simply inanimate matter which we can interact with and view objectively. They act on us, just as we act on them. This is one way in which matter matters, the way in which materials, living and non-living, directly impact on us. However, we are also now more aware that matter can have an indirect impact on us; the rubbish we discard, the food that is wasted, the plastics that reach the oceans, the components in our mobile phones that result in a poisoned Earth, the matter that we do not see at the other side of the planet. All that connects and comes back to us.

There is, as yet, no perfectly formed pedagogy which guarantees a new way of thinking about our relationship with the planet that we depend upon. What is certain, though, is that we cannot continue as before, and we must actively explore the alternatives from pre-school through to secondary and tertiary education. I am, however, in no doubt that this involves connecting with our natural world in a way that the current education system does not allow. In this essay I have touched on some ideas that we need to take account of in moving science education, and education in general, forward into a new ecologically aware education. One which is driven not by economic agendas but rather by ecological ones that take account of the overall health and wellbeing of children and young people as well as of the planet. Economics will thrive on a healthy planet with healthy people, but it cannot survive if the planet dies.

 *Donald Gray is a Professor in the School of Education at Aberdeen University. He has a particular interest in science and sustainability issues, Goethean science, STEAM and outdoor learning. He has been involved in education for over 40 years, first as a science teacher and curriculum development officer prior to entering higher education as a researcher and lecturer. He is currently involved in developing the idea of Garden Schools with the organisation One Seed Forward.*

Moral Intuition

EDUCATION SHOULD BE THE EXPLORATION OF SHARED VALUE SYSTEMS

Sacha Peers

> *"The aim of education is the knowledge,*
> *not of facts, but of values."*
> – Williams S. Burroughs

The attainment of 'good' A level grades at the end of last summer felt strangely anticlimactic. I had reached the end of formal education and was left with the feeling, 'What next...?'

Education is increasingly treated as a 'means' to other 'ends,' rather than an 'end' in and of itself. It is treated as the means to the end of good exam results which ultimately serve as little more than a signalling device to universities. This over-emphasis on exam results obscures what should be the real end of education: instilling a love of learning, cooperation, and communal exploration of shared value systems. It is the restoration of these shared values that are so crucial in rebalancing our relationship with the planet and within society.

Reflecting on my formal education, what I remember most fondly are class debates and discussions. Although I did not realise this at the time, they served to communally explore and challenge each other's moral intuitions; a process which is central to forming the shared value systems that bind society together. As each generation grows up, they inherit traditions, values, and systems of power from previous generations. Thus, the process of growing up must include opportunities to reflect on these values and to decide to what extent they should or should not be collectively reproduced.

Economics has long purported to be a 'value neutral' science, which has no 'end'; economic value was taken to be synonymous with social value. However, the economic system does of course have values, it does have an 'end'. The values of our economic system are utilitarian values of maximising efficiency. These utilitarian values of efficiency underpin our economic system and have produced corrosive effects, where market values crowd out social values such as protecting the planet and society. However, what our education systems do not teach us is that these values are collectively chosen and can collectively be replaced.

Upcoming generations, therefore, must be presented with educational opportunities to form values collectively; a process of considering and scrutinising the values that are currently inherent in society. For example, our education system should ask, 'Is the destruction of the planet and the creation of vast social inequality in the name of free market 'efficiency' something that students wish to replicate in the future?' The schooling system in its current form acts as a distraction from this crucial process of value reflection. Instead, students are ever presented with the pressure of the next exam, the next hoop to jump through, so much so that by the time they have completed their education, they have unwittingly absorbed and reproduced the values of the status quo.

> Our education system should ask,
> 'Is the destruction of the planet and
> the creation of vast social inequality in the
> name of free market 'efficiency' something
> that students wish to replicate in the future?'

Socrates and Aristotle both subscribed to a holistic view of politics in which education was treated as the central question. For Aristotle, the identity of a *polis* is not constituted by its walls; a city is not merely a defensive or commercial alliance, it consists of *philia* – a civic reflection of the familial bonds of loyalty and camaraderie that consist of more than mutual convenience. Citizens are those who share a common way of life, common memories, and common values. Aristotle asserts that all communities aim at some good, *ergo* the educational community should reflect these civic bonds of common good.

However, emphasis on civic education was lost in the West during the Enlightenment, which instead established the liberal tradition of individual rights before collective duties, a reversal from the ancient emphasis on the collective. The creation of rights of the individual, laid the basis for modern liberalism which is supposedly 'value neutral', leaving the individual to decide their own values – their own conception of what is good. While this is an intuitively attractive concept of human freedom, after reading Michael Sandel's writings on justice, ethics, and democracy, I believe its application has been fundamentally flawed and is corrosive to society and our relationship with the planet. We must look to our education system to restore emphasis on a values-based system of morality, one which has long been side-lined by the rights-based liberal order.

I would go as far as saying that a rights- and consent-based society is fundamentally morally unsatisfying. The liberal presentation of the State, which pits the rights of individuals against each other, presents a hollow vision of a society which is bound together by nothing more than self-interest. It is centred around John Stuart Mill's 'Harm Principle' – the idea that individuals should be free to act in any way that does not 'harm another' (whether this free will should be extended to 'harm of the planet' is not even debated). While on the face of it this is an intuitively attractive idea, there are undoubtedly examples of moral obligations that do not arise from an act of consent. Take for example the moral bond between members of the same family. I never 'consented' to having my parents, however, most people would agree that I have a moral obligation to look after my parents in their old age. This is one example of a morality that transcends the narrow bounds of consent and reflects something deeper about human nature and morality, something that is obscured when we talk about morality purely in terms of rights and consent. Similarly, the moral bond between citizens of a society that Aristotle referenced lies deeper than aligned self-interest and consent. Instead of pitting individuals' rights against one another, social wellbeing depends upon cohesion and solidarity – a shared value system. Our education system should prioritise, teach, and explore this shared value system which is ultimately the unseen force that binds a society together.

Unfortunately, individualism remains the ideology promulgated by our education system. Meritocratic individualism is the dominant ideology of the modern school. Students must vie with one another in endless competition, and the purpose of education is emphasised on individual terms rather than on social terms. This prioritisation of individual goals has come at the cost of the detriment of exploration and nurture of social values in the crucial adolescent years. However, it is a false dichotomy to suppose that one must either prioritise the individual or the social goals of education. Socrates recognises this in *The Republic*, when he draws an analogy between the city and the soul, by saying that justice in the city was only achieved through cultivating harmony between the competing forces within the soul, and vice versa.

I believe the parity between the individual and the collective has struck a healthier balance in Finland. The Finnish education system is consistently ranked at the top of international comparisons of schooling systems, as well as the country ranking first in the UN 2020 World Happiness Report. Finland achieves this high level of education while spending roughly the same amount per pupil as the UK. The difference is in their approach to learning and the perceived 'ends' of education. "We prepare children to learn how to learn, not how to take a test," said Pasi Sahlberg, former Director General of Finland's Ministry of Education

The lack of a centralised exam system – students instead being assessed by teachers – gives a high degree of autonomy to individual teachers and schools, allowing for a more engaging education, focused on promoting prosocial behaviour, creative thinking, and a love of learning. Perhaps the biggest difference between Finland's education systems and the UK's is the social value placed on teachers. This social value is independent of salary - which is roughly similar in the two countries. Instead, social value stems from a social recognition of the importance of education in society. All teachers require at least a Masters' degree, and competition for these places is fierce – in 2015 only 7% of applicants in Helsinki were accepted for the five-year Masters' degree to become a primary school teacher.

Finland provides a model for education to move towards, a model that places more emphasis on the social and civic aspects of education and attempts to balance them appropriately with individual aspects, such as learning how to learn. The Finnish system corrects the false dichotomy in the West – the choice of prioritising the individual vs. prioritising society. Socrates in his city-soul analogy saw no such dichotomy because he believed that to be socially fulfilled is to be individually fulfilled.

I am by no means suggesting that the Enlightenment project of rights is altogether undesirable, merely that it has become over-relied upon and requires balancing with the ancient value systems as recognised by Socrates and Aristotle. The value systems that will bind society together and rebalance our relationship with the planet are only rediscovered and adapted through education.

Sacha Peers is a philosophy student living in London with a deep passion for education and Nature.

Our Interconnected Oneness

EDUCATION MUST HONOUR THE MULTIFACETED NATURE OF INTELLIGENCE

Opeyemi Adewale

"Knowing yourself is the beginning of all wisdom"
– Aristotle

'We are all related, we are all one.' This succinct yet profound statement speaks to the rich kinship shared by all species on the planet, but this beautiful relationship is perpetually threatened by unwavering ecological negligence. With the outbreak of the coronavirus pandemic in 2020, humanity was confronted with an existential peril. Though it was an unspeakable tragedy, the events of the year, nevertheless accentuated the insight of this phrase. As the virus wreathed the world, its unheeded wisdom seemed to echo in the background, reminding us of something we have ignored as a species for so long – we are all interconnected.

The interconnectivity in Nature is intricate and exquisite, and working with Nature, rather than against it, is the highest degree of enlightened self-interest. Conversely, afflicting Nature is the crowning act of self-affliction. The pandemic proves it. Scientific data shows that 60% of all existing infectious diseases and 75% of all emerging infectious diseases are of zoonotic origin – meaning they come from animals. This occurs primarily because the natural habitats of non-human animals are constantly strained by human incursion in the form of continuous deforestation, poaching and wildlife trafficking. Anthropogenic climate change is more proof: it has caused approximately 1.0°C of global warming above the levels of the pre-industrial era (1850-1900). Simply put, the world is the hottest it has ever been in the history of recorded data. The impact of this includes, amongst other things, heatwaves, rising sea levels, desertification, and the loss of biodiversity. All these, in turn, threaten food, water, economic and regional security. We can no longer ignore the fact that the welfare and health of humanity cannot be divorced from the welfare and health of the planet. Therefore, the welfare of the planet should be the nucleus of all societal and communal action. It should be the *raison d'etre* for culture, politics, business and certainly education.

Modern education is a process of acquiring knowledge by receiving information from teachers. This 'cognitive worldview' aligns with the traditional precepts of education, which mainly involves the rigid training of the mind. Learners are flooded with information, and they respond by building dams, mentally. These dams retain information with the aim of completing a syllabus and then the dams are released and what has been stored is regurgitated in examinations. The objective is normally the acquisition of a certification which makes job procurement easier. General learning methods include memorisation, recollection, and reproduction.

This worldview manufactures workers who compete for jobs in a mechanised society, which as we now know, contributes to climate chaos, over-consumption, and the loss of Indigenous knowledge.

By contrast, the 'ecological worldview' of education prescribes that education should be a process of self-discovery in a viable organic environment. The educator is a coach who guides and 'leads-out' the learner. An instance of this is the kindergarten system founded by the educator, Friedrich Froebel. Kindergarten means the 'children's garden' and the system promotes the concept of self-activity, mostly Nature-based. The concept encourages children to freely explore their interests, discover them and be led by them. The ecological worldview asserts that education should be the building of partnership and relationships - between the learner and the educator, and between the learner and their environment. It encourages learning with the whole being and not only the cognitive faculty. Learning with the whole being is the holistic approach to education.

Education is an indispensable tool to combat climate change. Indeed, an 'inclusive and equitable quality education for all' is one of the Sustainable Development Goals (SDGs) which has a strong nexus with all the other goals – because it is clear that any effort made to accomplish the SDGs is futile without the thrust of holistic education. However, students still need more than robust cognitive learning and an ecological worldview; they need practical skills to transition seamlessly from theoretical work to real world practice. Also, social, psychological, and emotional learning is of paramount importance. The Brookings Institute states, "...young people need both a strong knowledge base around the causes of a warming climate [and] a strong 'set of skills' that will allow them to apply their knowledge in the real world, including problem-solving, critical thinking, teamwork, coping with uncertainty, empathy, and negotiation." The United Nations Children's Fund (UNICEF) calls these 'life skills' and defines them as a 'behaviour development approach' designed to balance three aspects of learning: knowledge, attitude and skills. According to them, life skills can be used in many areas including in consumer education, environmental education, and education for development amongst others. In addition, life skills include effective communication skills, decision-making, creative thinking, interpersonal relationship skills, self-awareness building skills and emotions.

The sad fact is that whilst over 50% of students globally are taught environmental education in formal education settings such as primary, secondary, or tertiary education levels, it is still cognitive learning whereas behavioural learning is almost non-existent, especially at the tertiary level. This means that globally, at all levels of education, cognitive learning is the preferred method of teaching environmental education. Social, emotional, and behavioural learning rarely receives attention and thus education is delivered without a holistic perception of the learner. This has an effect on the socio-emotional growth of the learner because it attenuates the benefits of environmental education, and we see the consequences in the multifaceted crises that now imperil life on Earth.

Holistic education is an inimitable tool for changing a world buffeted by ecological negligence, environmental crises and pedagogical deficiencies

Holistic education, on the other hand, appreciates the whole person and focuses on training all aspects of their being by developing, amongst other things, a learner's mental, emotional, artistic and spiritual aptitudes thereby creating a diversity of intelligences. When an educator 'leads out' a learner by encouraging them to experience their multifaceted intelligences, they often discover latent abilities and skills they didn't know existed, and these can become life-long channels of creativity and passion. The diversity of intelligences that are incorporated in holistic education are:

MENTAL INTELLIGENCE – creativity, cognitive aptitude, learning, critical thinking, intellectual power, imagination, originality

EMOTIONAL INTELLIGENCE – love, reverence, empathy, compassion, gratitude

ARTISTIC INTELLICENCE – aesthetic appreciation, crafting, building, weaving, arts, sculpting

SPIRITUAL INTELLIGENCE – altruism, self-sacrifice, graciousness, faith, kindness, goodwill

SOCIAL INTELLIGENCE – communicative confidence, relationships, persuasion, active listening, manners, teamwork

ECOLOGICAL INTELLIGENCE – eco-centrism, sustainability, environmental awareness and competence, ecological thinking

PHYSICAL INTELLIGENCE – motor skills, athleticism

PSYCHOLOGICAL INTELLIGENCE – fortitude, grit, valour, perseverance, self-confidence, motivation

CULTURAL INTELLIGENCE – cross cultural mindfulness, racial sensitivity, tolerance, respect, fairness

MORAL INTELLIGENCE – Moral courage, discernment, good judgment, values, honour, virtue

Holistic education is an inimitable tool for changing a world buffeted by ecological negligence, environmental crises, and pedagogical deficiencies. A global desire for climate action cannot be separated from holistic education underpinned by the principles of ecology. Holistic education moulds an emotionally balanced, ecologically intelligent, and intellectually inquisitive learner through a process of self-activity, self-discovery and self-exploration redounding to myriad positive outcomes in all facets of the learner's life. This will inspire and build an ecologically perceptive world where all can harmoniously dwell.

 *Opeyemi Adewale was born in Nigeria. He has a Masters' degree in architecture and is a two-time semi-finalist and a finalist in the prestigious University of Berkeley Prize for Architectural Design Excellence. He is also the recipient of the special prize in the United Nations Convention to Combat Desertification (UNCCD) global essay research contest 2021. He currently resides in Lagos.*

The Warp and Weft of Ethics

IT IS CRUCIAL TO WEAVE VALUES-BASED DIVERGENT THINKING INTO EDUCATION

Shakti Saran

*"Educating the mind without educating
the heart is no education at all."*
– [Att.] Aristotle

In early 2018, I had the good fortune of attending a summer camp organised for tribal Children near Hyderabad, India, sponsored by the local government. Although the notion of education, for most people conjures up images of a blind paper chase, a swirl of classrooms, textbooks, tutorials, and high-stress examinations, what I saw instead was children engaged in improvisation, visualisations, Nature walks and more. I realised during this visit that education can equally be a process of observation, discovery, self-learning, a fun-filled connection with Nature and.

Just a few weeks prior, as a part of a leadership development programme for those aspiring to work in the social sector, the views of a fellow participant stirred something in me. He stated, "There are many non-governmental organisations in the education sector.

Most or all of them are focused on getting basic language and math skills fixed. Are there any initiatives in India where education is also encouraging children to think critically, question social practices and help them grow with a more humanistic mindset? Otherwise, aren't all these efforts feeding into the same social and economic systems by providing more foot soldiers? The very same systems which the development sector is questioning?" My visit to Hyderabad a few weeks later answered the questions that my colleague raised.

Before addressing the necessary transformation of education, it is imperative that we understand what education actually is. Education means different things to different people. In a conventional sense, and in the way the United Nations addresses it, education refers to the process of learning from early childhood to primary, secondary, tertiary and even adult learning. Sustainable Development Goal (SDG) No. 4 focuses on the goals of education which extend to technical, vocational training and skills for work.

As laudatory as UN SDG4 might be, it does not *fully* address the case for values in education. The late Anthony de Mello, a Jesuit priest from India said it eloquently: "There are two educations: the one that teaches how to make a living and the one that teaches how to live." In his essay 'The Greatest Resource: Education', E. F. Schumacher answers the question 'What is education?' by saying, "It is the transmission of ideas which enables humanity to choose between one thing and another or to live a life which is something above meaningless tragedy or inward disgrace." Schumacher goes on to speak about how essential it is to have a metaphysical core in education and ensuring the light of consciousness falls on it.

Observe the contrast between these conventional and unconventional descriptions of what education is. I would say there is also a third way, which brings both in balance. Literacy and numeracy are valuable skills and stepping-stones to poverty alleviation. A patient undergoing surgery would be at risk if operated on by a good-hearted but unskilled surgeon. Yet, it is hard to reconcile the anomalies we find in the educated person. Educated people are largely responsible for the abysmal state of our planet. For instance, they contribute more to pollution than uneducated people. The way history is taught in schools is often a basis for perpetuating racial inequalities. If you study financial crises that have rocked the globe over the last three decades you will find all of them having been precipitated by highly educated people, some of them from leading universities in the world.

Conventional education is only a starting point. Educated people as the source of our problems raises the issue of the role of ethics in education or more so, the lack of it. Morality or ethics, in its raw form is unpalatable unless one is a student of Divinity or Philosophy. Unfortunately, most of our education systems are long on convergent thinking, which focuses on a single path to a single answer, and short on divergent thinking which presents itself as multiple paths to multiple answers. How do we then weave ethics into our education system without people raising their defences?

Educated people as the source of our
problems raises the issue of the role of
ethics in education or more so, the lack of it

To move forward, the third way in education
brings convergent and divergent dimensions
together to create a binding synthesis with universal
values and ethics embodied at the core of education.
Values-based education and the natural sciences,
engineering or management are not mutually
exclusive. The more woven and integrated these are,
the more people will be receptive to them. Education
systems that are oriented towards divergent thinking
are naturally capable of imparting ethics. The pill of
ethics, which needs a right-brain disposition, can be
swallowed easily with liberal and experiential forms
of education.

There are several innovative ways in which education
can be designed and transformed as if people and
planet matter. In many parts of the world curricula
are structured rigidly. It is essential that students be
given a choice or a range of electives to pursue their
natural preferences as witnessed with the new National
Education Policy in India. Education should promote
natural learning and not be lopsidedly focused on
academics and examinations. Academic study should
be at best half of the learning experience and the rest of
learning should be experiential.

Introducing divergent approaches, in classroom study can bring convergent aspects into balance. For instance, high school curricula should have mandatory courses in preventive health or environmental regeneration. Management students should be encouraged to enrol in a case-study oriented course on ethics and every degree in economics should have embedded courses in ecology.

Humanistic education calls for a significant enhancement of the experiential dimension that is currently missing. Experiential education triggers a shift from studying to learning, from examinations to discovery and supplements academics with astounding results. Rabindranath Tagore set up Santiniketan where students were taught outdoors in the lap of Nature. In the new education paradigm, learning from field visits and being in the outdoors needs to be stepped-up. Students should be coached in storytelling and story-boarding skills, outdoor survival, Nature appreciation, first aid classes, drama, Model UN, and more. Student exchanges like the reconciliation camps organised by Seeds of Peace need to be encouraged. Students, be they from medical, engineering or management backgrounds, need to be motivated and incentivised to support social sector entities, in addition to the private sector, as part of summer internships.

Schumacher stresses that education needs to be a form of awakening and ought to lead to a higher level of 'being,' and also be the vehicle for the transmission of values and ideas that lead to the inner development of humankind. My visit to the summer camp near Hyderabad, taught me that more than academic fluency, we need our children to acquire crucial 21st century skills of communication, collaboration, and critical thinking. When we think in terms of experiential education, we are not limited by classroom nor age. Because education is essentially a life-long experience.

Shakti Saran is a former management consultant and corporate sector executive who crossed over to the social sector in 2017. He works as a Senior Fellow at PYXERA Global India. On joining the non-profit sector, he quickly realised that aspiring for a thriving humanity and planet calls for a systemic view of life and has since been championing the case for addressing complex global challenges through a systems lens. He is an alumnus of India Leaders for Social Sector, Radical Transformation Leadership and Capra Course and has a certification in systems thinking from Cornell University.

Spiritual Education

TOWARDS A SENSE OF THE SACRED

Guillem Ferrer

*"We are not human beings having a spiritual experience,
we are spiritual beings having a human experience."*
– Wayne W. Dyer

All humans without exception are spiritual beings; we are pilgrims on the path to our final destination, which is to discover within ourselves, the presence of the Universal Being, the Universal Creative Intelligence that gives life to all and unites us as one family: the Earth Community. The Universal Being is the Supreme, the Eternal, the Truth of the Universe, a state of consciousness that is beyond time. In humans, the Universal Being is our soul.

The wise say that in everything there are two dimensions: the visible dimension and the invisible dimension. Matter, that which we can measure, is the kingdom of the Earth; and the soul, that which we feel, is the kingdom of Light: this invisible dimension cannot be understood purely by reason – in fact, it can only be grasped through spiritual imagination, through intuition, through that which makes our heart leap.

What is the point of conquering the material world only to lose our soul? Are we prepared to become aware of our true Being? To live and share spirituality and ecology as sacred activists? The seed of spiritual wisdom is already within us. A good teacher touches that seed allowing it to awaken, sprout and grow. There is no knowledge higher or more conducive to full satisfaction than knowing one's own Self. We are here to awaken into this awareness. Let children self-design their own schooling and learn to connect to their innate wisdom, the wisdom of their hands, of the Earth, of community, and thereby discover the great mystery: 'I am.'

Our children have been born into a new era, so let's not limit them to what family, teachers, schools, and universities know; let us walk alongside them with our presence, our wisdom, and our creative love. Children carry the 'great teacher' inside them, deep within their Being. Just as the seed knows how to become a tree, and the bird knows how to build the perfect nest, the child knows how to access their innate wisdom. The forces that guide the stars are the same forces that make our hearts beat; the Universal Intelligence that awakens each Spring is the same intelligence that awakens our dreams. The sun, the stars, the mountains... they know – and we know. Everything is connected to the Creative Universal Intelligence.

If you want to understand this Intelligence, follow Saint Augustine's advice: do not search outside, because the truth lies within you. Listen to your consciousness, since everything else is conditioned by it, beginning with the way you relate to yourself and then to other living beings, to Nature, to the material world. You, child, are the teacher. The Creative Intelligence of the Universe will guide you. Let us learn from the invisible, from the sacred dimension of life. Knowing who you are, knowing yourself, is the highest offering each of us can make to the whole of humanity. That is the essence.

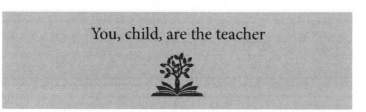

You, child, are the teacher

Educating our Inner Being offers us the opportunity to reconnect with the spiritual and ecological dimension that lies within each of us. Wisdom should not be confused with intelligence, information nor even with education. You may be highly educated yet show a complete lack of wisdom, or you can be illiterate yet be connected to that inner kingdom of wisdom, which is the dimension of consciousness or presence. So, the most important thing to learn is how to be in contact with this inner dimension of consciousness. It's not just another subject to add to the academic curriculum; it should be the foundation of all teaching in school and in life itself.

Our relationships with people and Nature are based on the quality of our Being. When that quality has an inner stillness, the terrain upon which future actions occur will change and thus events will take a different turn. We simply return to our true home, reconnect with ourselves, with our own spirituality. Our most significant interaction is with our true nature. In this way we offer harmony to those with whom we connect. Connecting to one's limitless qualities in daily life, to then connect in the same way with others and with the world, is surely the path towards the realisation of our Spiritual Being.

The wisdom of spiritual education is that it is based on commitment to truth and non-violence. It teaches how to express love in our daily actions, through a method based on ethics and work. Spiritual education moves outwards to the hands and the senses, to the brain and the heart; to school and to society, all the way to the Universe.

Creative love is the way to turn life into light, it is the interface between the visible and the invisible, the way to reach the Universal Being. Socrates tells us that, "Love is the messenger between the visible and the invisible," while St. John of the Cross affirms, "We can only call love that which connects the soul with the Universal Being." Saint Teresa explained that the "important thing is not to think much, but to love much," and that, "The love of the Universal Being should not be built in our imagination, it has to be experienced. We experience love in contemplation, in the silence of the mind, when we enter the spheres of poetry and art, where true love and joy can be found."

Love, invisible and immeasurable, is the force that moves the Universe. Just as the rational mind can see that all matter is energy, so the soul can see that all energy is love; any part of creation can be a mathematical equation for the mind and a love song for the soul. Love leads us to light. It is not enough to have more nor to know more – we need only live more, and if we want to live more, we must love more. Love is the foundation of spiritual education.

Educating the Inner Being awakens people's consciousness and the ecology of Nature; it protects and regenerates the planet's ecosystems and delivers a message, a clear mantra for all of humanity that is precise and precious: all is one, all is connected, inter-related and inter-dependent. We are a large family sharing planet Earth.

If we cultivate consciousness and recover the spirit of the sacred, our egos will surrender and we will discover the beauty, goodness, and truth in the coexistence of the animal, the vegetable, the mineral, the human and the divine kingdoms. Together they make up a whole within the laws of Nature and of the cosmos. Spiritual education is the path of wisdom, of love, of light and of the Universal Being.

Being a child
is believing in love, believing in beauty, believing in kindness, believing in faith;
it is being as humble as the Earth, as innocent as a butterfly, so small that the ants can whisper in your ear;
it is living in an almond shell and feeling like the queen of infinite space;
turning pumpkins into carriages and mice into horses; it is transforming the small into the sublime and nothing into something; because every child has a little fairy in their soul.

Being a child, in William Blake's words is:
To see the world in a grain of sand
And a heaven in a wild flower
Hold Infinity in the palm of your hand
And Eternity in an hour.

 *Guillem Ferrer is a Peaceful Activist. in 1998 he founded and is now director of PocaPoc, a movement of activists that inspires, encourages and creates local actions for caring for the Earth, the soul and society; and of the Education for Life Foundation, helping to give birth to a new holistic education that prioritises self-knowledge and self-sufficiency.*

IMAGINATION ENCIRCLES THE WORLD
WE NEED TO RETHINK THE WAY WE THINK

Sarah Wilkinson

"Imagination is the source of every form of human achievement. And it's the one thing that I believe we are systematically jeopardizing in the way we educate our children and ourselves."
– Sir Ken Robinson

"Don't cast a clout 'til May's out," my grandmother, Elsie, used to say. I learned the hard – or rather, cold – way to remember her wise words when walking without a jacket on an unseasonably warm April day only to find once the clouds set in that there was a distinct chill in the air. This gift that Grandma Elsie gave me –presumably given to her by a parent – will no doubt be passed down the generations of my family. Grandma Elsie would not have been considered an 'educated' or particularly notable person in society – she had various part time jobs, raised three children, and kept a clean, tidy, welcoming home. But the values that she handed down to my mother (some of which were then handed down to me) are priceless. She was filled with everything that is good – kindness, generosity, wisdom, and honesty. Today, these core values are being overshadowed by the unhealthily competitive nature of mainstream academia, sadly bringing about more unfulfilled desires than ever before, breeding an incessant, insatiable hunger that completely neglects the most basic human needs, and therefore the stability of mental health, along with the environment we inhabit.

How do humans pass on values to their children? As history informs us, people have always told each other stories: true stories, made-up stories, and allegoric stories. But all stories, like trees, are rooted somewhere. Today we live in a society that neglects the 'imaginary world' and places more importance in something that is referred to as the 'real world,' but as Einstein is believed to have said, "Knowledge is limited. Imagination encircles the world." If we are losing the ability to tell a story, how can we hand down our core values to our children? In my family, (for whom there is no barrier between these two places) the characters and stories that we create and the things they teach us are as real as the grass we walk on.

The potential of the imagination is infinite. Doesn't everything we know about come, in one way or another, from a story? What changes occur to make that story become fact or knowledge? The consistent thing that has inspired human stories throughout history is Nature. Immersing children from a young age in the natural world may be part of the solution to the problem of preserving our home planet and all its beautiful residents. Yet more and more, the way children are educated is leading them away from their natural instincts, particularly the instinct to be outside in Nature, and from their inherited values, and further towards conformity as a cog in a machine that is destroying our biosphere. We are simply not listening. Children are growing up believing that they cannot be trusted – a message reiterated to them throughout school and into their working lives. Over time, we have placed less value in the thoughts that young people form for themselves through self-discovery and added infinite amounts of value to lists of facts that they have been taught. Therefore, there is less value applied to how young people think and learn, and more value given to what they apparently need to succeed.

We need to re-find value in our values. Because our beautiful home planet matters, along with every living thing. It matters more than the realms of our modern society convince us to believe. Grandma Elsie mattered; she made a home, she re-used, she 'made do' and she knew how to look after those close to her. We are now raising list-making, fact-remembering children in a society where it is cheaper to buy a colourful multipack of synthetic potato crisps full of salt and unhealthy fats than a punnet of strawberries, where pre-cooked frozen baked potatoes are sold in microwaveable plastic boxes, where care homes for old people are forever growing and disconnecting families, where we have forgotten how to fix, repair and be content with what we have. We need to re-think the way we think. We need to establish a new world where every single living thing has infinite value; a natural, rewilded world that we don't simply continue to take from, but that we work in and live alongside in natural harmony. But how?

As Alastair McIntosh said, in his book, *Riders on the Storm*, "In opening up the realm of metavalues...

it helps us not to miss the difference that can make the difference." Our children are that difference. It should be of the utmost importance that a child retains their sense of wonderment, and that their light can continue to shine. Children are born without prejudice; they are born unknowing of the order that has been created for them to fit into. And they are born unaware that if they cannot or will not fit into that order then they may be deemed unfit for purpose. The changes are within our grasp; the responsibility to guide our young people as they navigate through life lies in our hands. It cannot be said any more eloquently than in the words of the Jedi master, Yoda: "Pass on what you have learned. Strength, mastery... but weakness, folly, failure also. Yes: failure, most of all. The greatest teacher, failure is."

If we let our young people lead, and actively listen to what they have to say without our fears or ego overcoming us, if we let them stand on our shoulders to see what is beyond, then they will flourish and grow into what the future demands of them. Bill, who was a dear friend of my husband's, use to say, "Questions open doors. Answers close them." We need to let our children's questions open a thousand doors of exploration, to anywhere, anyhow, with any tools they need. Not close those doors because it's not in the curriculum.

> If we let our young people lead, and actively listen to what they have to say without our fears or ego overcoming us, then they will flourish and grow into what the future demands of them

Let us release the constraints on children and let them decide. Let them back into the real world – not the 'real world' that has been created for them, which is sending them into an uncertain future, along with growing mental health problems and filled with the mess those before them have left – but the real world of Nature and the imagination. Let us not lead them into a site so littered with destruction that they can no longer see the grass; where everything is burned to the ground, but let our young people tell us new stories of the future they imagine.

Sarah Wilkinson is a musician, maker and writer. She was a music lecturer and tutor for 13 years, and now runs her own small businesses. Sarah is an unschooling parent, and a passionate advocate for change in the education system.

Fruitful Questions

to ask, 'what is true?'

Dana Littlepage Smith

"A well-educated mind will always have more questions than answers."
– Helen Keller

Fourteen years ago, on the eve of the Stern Review of Economics and Climate Change, I sat in a classroom asking students to write essays on the care-filled stewardship of our planet. One child replied, "We are all idiots! If it is true that we are committing ecocide, why are we sitting here writing essays? Why aren't we walking, walking all day and all night into London to ask our government and churches to act?"

Sam's question shaped the course of our learning. Indeed, a small group of students walked out of the classroom and together with some of their parents, plodded along 70 miles of the Pilgrims' Way from Canterbury to London where Rowan Williams met and listened to their questions and concerns. After delivering their questions to Downing Street, they then questioned their MP. The police joined us to question our gathering. This spurred more conversation, and these questions became the core of our learning. As we walked, another student asked, "Why aren't more of our classrooms outside?" What a good question: the night sky taught the first navigators, the Phoenicians and Greeks. A ten-year old student also asked a simple question. "Is it true that writing essays is the most creative response to the destruction of our earth?"

The purpose of education is not to give answers but to train us in asking fruitful questions. That old stone cutter in Athens, intent on liberating the youth of his city believed the question, 'Is this true?' was the basis of education. Wisdom, after all, may simply be educated ignorance. Socrates The Wise proclaimed, "I know that I do not know!"

Socrates and Plato were deeply suspicious of the new art of writing; they understood its dangers. Writing will lull us into a false belief that we know things. Sophisticated cultures from the Sioux to the Kalahari Bushmen living sustainably within complex ecosystems have never relied on this form of transmission for learning. External marks on a page, even if I make them, do not imply wisdom. Oceans of information at my fingertips mean nothing if I cannot interrogate, integrate, analyse, understand, and make them my own. Our testing systems have little to do with empowering humans to become wise in the face of today's challenges.

If we had not listened to Sam, a living element of our learning would have died. Humans are made for and by connections. Academies were born when Plato walked into a field owned by a man named Akademos, taking with him his teacher's idea that we learn by engaging in heartfelt, free-wheeling conversations. Their favourite question: 'Is this true?'

Our world is experiencing exponential change: we as a species are linked up and making radical leaps, so why are our learners shutting down and turning off? What is it that our children want and need in their learning? Education always begins by asking questions.

A young man I tutored was cramming for his A-levels; he had spent hours copying out answers verbatim that would win him marks. He is smart and knows what is expected of him, yet on the day of his exams after 20 minutes, he walked out of the hall. He was predicted to get straight A's and his father was enraged. When I asked the student what happened, he replied, "This is all so meaningless. I would rather be a hunter-gatherer." Then he spilled out a slew of questions to really hammer his point home: "When we speed past the moment of singularity, will I be happy, will we even be here as a species? Who gives a monkeys', if I'm a rich bastard, but have to live with armed guards because climate refugees are desperate for refuge on this island?"

If to educate is to encourage the voice within to ask, 'What is true?' throughout our lifetime, then he is on the road to a glorious learning adventure. I tell his father as much, but he wants his son to get a good job. Cultivating a creative habit of mind which forges connections is, in the father's opinion, not the point. Perhaps students' disconnection, depression, anxiety, trauma and despair are not the point, either? But, of course, this is exactly what is true.

Even if the questions of our students too honest, painful and unwieldy, even if the questions are too connected with our own pain, we will lose our students if we do not embrace their questions and champion their spirit. Their spark is still alive, but the flint of the educational system can only fail them so many times before it dies. I listen to them; I need not know the answers. As the poet Wislawa Syzmborska wrote in her Nobel Acceptance speech, 'There are four words which can take us anywhere; they rise on mighty wings. *I do not know*". In our unknowing, might we devise new forms of classrooms and learning?

Any day we can witness miracles in
uncounted classrooms across the planet

John, a committed and visionary teacher in a farming community in North Devon believes that the freed mind is its own teacher. He believes children can learn how to be their own best teachers. He scrapped the old syllabus and encouraged children to discover their own effective modes of learning. Could they build a plane in order to question their ability to read with precision and apply their learning? Might they investigate the laws of aerodynamics in real-time? Would the board of governors trust the students' illimitable potential and fly their machine? John continues to encourage and dare all of us as learners and educators. Can we invite students from rural, under-supported schools in North Devon to produce and enact Henry V in 48 hours? Will their parents and grandparents flock to a field to witness the

Battle of Agincourt where their children-actors have memorised great swathes of the bard? Yes, is the answer to all of the above. Any day we can witness miracles in uncounted classrooms across the planet.

After being incarcerated in Auschwitz, Viktor Frankl noted that the place of liberation lies in our ability to pause, to reflect, to act and not simply to react. What happens in our classrooms should be holy, sanctified, because good education connects humans, all species, and our planetary home. But what price do we place on pausing in these moments to honour this connection? Dare we even slow down enough to see one another, our hearts connected to our thinking, in these times of such speed and stress and anxiety? It is possible to embrace our deepest humanity in every classroom in any place in the world, if we ask, 'What is True?' and honour the fact that we may not know. Even the blind Gloucester, whose eyes had been poked out, woke to a new reality when he spoke to his King, Lear: 'I can see. I see feelingly.'

Our students are our teachers; pandemics and the climate crisis are our teachers. How are we to live truthfully in a post-truth world? Truth bids me inquire within and alongside the timeless circle of learning: from the mathematician Brahmagupta, who created 'Zero' in 628 CE, to Ada Byron Lovelace who wrote the first programme for a computer. This essay is being written by countless hands and hearts and questioning minds. We can support one another by trusting that circle of learning which is interdependent and ever-expanding. It has moved through the Buddha, Black Elk, Greta and Malala... It lives by questions and paradoxes and in long ruminative silences. It leaps alongside ages of unknowing... It might be here, now, gathering momentum, within and between us.

 *Dana Littlepage Smith has been teaching for over 30 years. She has worked in various countries alongside a wide spectrum of learners including elders, Quakers, prisoners, children in orphanages, primary school children and university students. She is an American poet who has published five books, her latest being* What Love Requires. *She has lived in Devon for the last 25 years with her British husband.*

PART THREE

IN PRACTICE

Breathe Out

CAN ONE EDUCATOR MAKE A REAL DIFFERENCE?

YES!

Stephen Sterling

- Tap into the energy, passion, ideas, and concerns that young people and students voice and help these fire and sustain your own motivation, commitment, and learning.
- Trust and encourage learners to express what they already know, value, and feel, and to say what they see as relevant education. Help them value themselves.
- Recognise that good education has always been about the heart and hands, as well as the head, about the arts as well as the sciences, and imagination as well as knowledge.
- Value learning experiences that nurture such qualities as self-confidence, joy, play, caring, empathy, creativity, curiosity, boldness and agency at individual and group levels.
- Teach and practice ecoliteracy.
- Embrace rather than avoid ethical issues and dilemmas as ways in to understand real world complexity, multiple perspectives, and values.
- Question policies that you know are limiting your creativity and energies, and find ways to work around

such constraints in and out of your institution, through self-care and working with others.

- Inform yourself, and help your learners navigate the new digital world of misinformation and half-truths critically. Help them become critically active citizens for eco and social justice and alternative futures, rather than passive actors in current trends. It's their future at stake.

- Network and collaborate with as many allies or co-conspirators as you can - inside and outside your institution - to enliven the learning spaces and learning experiences: through inter-disciplinarity, working on real-world issues, using varied participative and transformative pedagogies, collaborative learning and co-inquiry, dialoguing, employing action research, using the arts, learning outdoors, and with the community. As far as you can, resist the 'delivery' and target culture and make learning experiential, participative and explorative – and meaningful.

- Celebrate successes and use them to help change the culture of the institution.

- Think about how to make 'everything' curriculum – as lived experience, with the institution as far as possible a living lab and exemplar striving towards sustainability.

- Find out about 'wild pedagogies', and ensure learners have immersive and reflective time in nature. Have them develop practical skills, through for example, growing plants and vegetables, making food, making green and beautiful spaces, encouraging wildlife etc. Cultivate appreciation, respect and love.

- Get into 'green competencies' and in particular systems thinking and critical thinking – particularly around interdependence, interrelationships, and agency – and bring these into the educational experience.
- Be inspired by – and have your learners engage with or begin – exciting regenerative initiatives in the community around food and growing, energy, health and wellbeing, a vibrant local economy, green transport, biodiversity and rewilding, the arts, participative democracy etc.
- As far as possible de-stress yourself and your learners and make regular time for reflection and enjoyment as a caring learning community.
- Start anywhere, however small, in creating a better present and future now – and maintain resolute hope. Believe in possibility.

The Small School

HOW TO NURTURE HEAD, HANDS AND HEART

Caroline Walker

*"It is of little use to call for a very big change in values
without at least incorporating those values in some
new structures, no matter how small."*
– E. F. Schumacher

In this essay, I will simply describe my experience of a
school, founded in 1982, where for 35 years, education as
if people and planet matter actually took place. Inspired
by E F Schumacher's *Small Is Beautiful,* the Small School,
an independent school in a Devon village, offered
pupils aged 11-16 small classes and a broad, balanced
curriculum of academic, creative, and practical activities
to nurture head, heart, and hands. The model was the
family, not the factory: no uniforms, of course; everyone
on first name terms; good relationships replacing rules
and regulations. There was no selection on grounds
of ability, or of ability to pay: the only criterion was
residence in the parish. Funding came from parental
contributions, charitable trusts, and individuals.
Personalised learning programmes were agreed between
teachers, parents, and pupils. All parents were invited

to monthly meetings and could serve on the governing body alongside community members. Parents with skills could teach or help in the office, and everyone joined in termly clean-ups and maintenance.

In a school of 1500 pupils, it is hard to make your voice heard or to feel like a co-creator of your learning: in a community of 36, each person is known as an individual and can express their views. The Small School met in a circle first thing every morning and at the end of each day to solve problems, resolve conflicts and plan together. Taking care of each other and taking care of the environment were the overarching principles, embodied in two foundational daily activities: cooking lunch together and cleaning the school together. Membership of the village community was also a vital part of pupils' development: the school (an old chapel) being permeable to the community, offering rehearsal space for music and drama groups, inviting adults into lessons, cleaning up footpaths and restoring woodland. Pupils were taught pottery by the local potter, woodwork by the local carpenter, art by a local artist. Older children had work experience in local businesses and farms. Village playing fields were used for games, local tradespeople were employed, and supplies for lunch bought in the village shops or grown in the school garden. Each child, on a rota basis, took turns to cook lunch. By the age of sixteen, every pupil, supervising two younger ones, had learned to plan, shop, cook, serve and wash-up a meal for 36 without adult help. In fact, many of the creative and practical activities, including gardening and building projects, had pupils working naturally in mixed-age groups. Gratifyingly, a number of former students even returned to teach.

> The model was the family, not the factory:
> no uniforms, of course; everyone on
> first name terms; good relationships
> replacing rules and regulations

Every summer, six 'Special Weeks' offered unusual and enriching activities – sculpture, creative writing, drama, science, music – allowing themed explorations and enquiry-based learning. Trips to the theatre in Plymouth and London, or abroad, for example to Denmark to participate in a charity project, to Italy, working on an organic farm, to Poland, on an exchange with an English language school, and to Japan, as guests of a mountain village where the school offered Medieval Mystery Plays and learned Kabuki theatre in return – all were memorable and profoundly educative experiences.

All this is possible in a human scale, community-based model of education, centred on the needs of the child, not of the examinations, still less of the economy. At the Small School, less academic pupils could still shine by providing a good lunch, making shelves for the classroom, helping a younger pupil, or organising a fund-raising event. In a small community you learn that you are responsible for your behaviour, because it affects others. You can have informal, respectful relationships with adults. Participating in, rather than passively consuming, education and taking care of your environment, inside and outside the school buildings, are enabled in human scale settings. This brings us to the wider question of: 'as if the planet matters.'

How is human scale education good for the planet? It starts with asking young people to clean their classrooms. Any mess you make, you and your classmates will have to clean up. At present schools and other educational institutions impart an unspoken but powerful lesson that at the end of the day, poorly-paid and almost invisible workers will come in and tidy up after you. This mentality, so pervasive we hardly question it, is the foundation of harmful and polluting behaviours: litter in the countryside, sofas fly-tipped in lay-bys, shopping trolleys pushed into canals; old computers shipped to Africa where children breathe toxic fumes extracting precious metals from them; plastics shipped to poor countries and destroying their environment; mines spilling poisons into watercourses and contaminating drinking water – all the result of people not learning to clean up after themselves, and it starts at school.

Care of the planet also starts with school lunch. Our current diet, mostly the product of industrial agribusiness, is implicated in five major challenges currently facing humanity: climate change, desertification and soil degradation, loss of biodiversity, zoonotic pandemics, and contamination of fresh water. It also makes us ill. Outsourcing of school catering to for-profit companies has led to a race to the bottom as far as quality and health benefits are concerned. Many children bring cheap processed food in lunch boxes and some schools even sell sugary carbonated drinks. Teaching children to cook using fresh ingredients is the first step towards raising awareness of food issues. If you require all children to cook, and if, as is very likely, at least some of them do not eat meat, then the meal should be vegetarian, to be enjoyed by all. As you cook together, you can debate intensive animal rearing, its multiple effects on the environment and its role in the outbreak of pandemics; as you grow vegetables, you can examine the benefits of organic farming on biodiversity; as you devise meals together you can discuss the effects of highly processed foods on health (and still make an occasional birthday cake). You empower young people to be more discerning in the way they source their food and give them a precious life skill in being able to cook for themselves and others.

It is customary, when advocating that all schools should have a kitchen and garden, to point out the mathematics and science 'attainment targets' achieved through cooking – how you can teach biology through growing vegetables, and chemistry through making bread from scratch; how you can even bring in some geography or social science. Yes, you can, if you wish to: that's education of the head; but it's more important, for the education of the heart, to sit down and enjoy the meal together, having a conversation and building community, and education of the hands comes through working with soil and other-than-human beings such as plants, worms, butterflies, fruit trees and compost.

I have painted you a picture of a school which set out to teach as if people and the planet matter. It is also a powerful example of the fact that size matters too. As Leopold Kohr said, "Wherever something is wrong, something is too big."

To align education with the needs of people and planet, we must re-examine the conflation of schooling with childcare. Most people cannot contemplate challenging current school systems, asking, "What would children do while parents work?" One brilliant solution comes from architect and designer Christopher Alexander in *A Pattern Language*, where he proposes in each neighbourhood a children's centre providing a kind of extended family. Alongside this, a network of learning mentors, each with responsibility for a small number of children, to facilitate and enable learning experiences – projects, thematic enquiries, personal explorations, in groups or as individuals. These activities take place in the community, using its resources: small 'shopfront schools', apprenticeships, a whole network of learning opportunities, enhanced (we can add in the 21st century), by online resources. In this way you replace 'the lock-step of compulsory schooling' (Alexander's term) with a system providing the necessary childcare for working parents alongside but not inextricable from learning opportunities that take full advantage of the wealth of knowledge and skills in the neighbourhood and online.

As Schumacher has pointed out, education is a divergent problem. There can never be one single answer to the question: 'What is the best education?' But in its current state, the education system shows no evidence of addressing emerging challenges: it resists change when the rest of society has changed around it. In the context of the current pandemic the traditional overcrowded and poorly ventilated classroom looks not just obsolete but positively dangerous. For the sake of people and planet we must, in David Fleming's words, "build lean, small scale, elegant, sustainable-resilient replacements" to the education system we have now. Can we seize this moment of turbulence and change?

 *Caroline Walker worked with development projects in a South Indian village before joining the Small School, Hartland, Devon, where she taught for 14 years, becoming head teacher from 1996 to 2000. Her MSc dissertation,* As If Size Mattered: Education for Sustainability and the Human Scale Approach *was published in 2008 by London South Bank University in their book* Journeys around Education for Sustainability.

Experiments in Education

WHERE COLLABORATION AND COMMUNITY, CURIOSITY AND COMPASSION ARE CRITICAL TO LEARNING

Alan Boldon

"The great shock of twentieth-century science has been that systems cannot be understood by analysis. The properties of the parts are not intrinsic properties but can only be understood within the context of the larger whole. Thus, the relationship between the parts and the whole has been reversed. In the systems approach, the properties of the parts can be understood only from the organisation of the whole."
– Fritjof Capra

In a time of climate emergency, obscene inequity and massive biodiversity loss, society is struggling to adapt. Whilst it is impressive that we can find vaccines at a rapid pace to tackle a pandemic, we are seemingly unable to answer the question of how and why the pandemic started. Multiple possible interlinked causes provide an unsettling challenge when we have such a preference for single, linear cause-and-effect narratives. We look for solutions in the form of quick fixes so that we don't have to worry about the causes, or otherwise we look away, as we seemingly career towards devastation, as if it is preferable to see how this disaster movie plays out rather than change the script.

Similarly, our learning institutions are struggling to adapt and reimagine their form; we may add some new subjects here and there in an attempt to equip pupils with a sound understanding of risk and the knowledge and skills necessary to make safe and informed decisions, but fundamental change is resisted. Our universities are structured into isolated subjects, schools and faculties that were established centuries ago. In their silos they analyse the world and break it into parts to understand it – and then look for synthesis, with clusters of disciplines occasionally brought together to combine perspectives, but we still do not interact with the dynamic whole.

In order to engage with the patterns, characteristics and properties of whole systems (such as education) we need to draw on many ways of knowing. This presents a great challenge when our ways of knowing have been busy staking out territory to differentiate their disciplines, but with the many interlinked crises facing the world we desperately need to find joined-up ways of knowing. Understanding how our worldviews have been constructed, including the vital work of rooting out deep prejudice, is a necessary first step towards the kind of wisdom and imagination that may help us to revitalise and reshape our relationships to each other and to all life around us.

At Dartington Trust, we are exploring our own worldviews in order to establish replicable forms of regenerative learning in service of people and planet.

We are reimagining and reshaping all our learning programmes and this process is embedded in a review of our culture, prejudices, habits, and institutional forms.

The founders of the Trust, Dorothy and Leonard Elmhirst, were deeply committed to learning by doing. Dorothy was central to the founding of The New School in New York. This re-invention of higher education was based partly on American philosopher and educator John Dewey's approach to problem-based experiential education and came soon after the tragedy of the first world war. It is often the case that times of crisis lead to creative responses and adaptation. Leonard and Dorothy, inspired by both Dewey and poet Rabindranath Tagore's commitment to learning that involves the 'head, hand and heart', established Dartington as a place for experiments in education. Other notable examples on the estate include Dartington College of Arts, Dartington School with its emphasis on experimental pedagogy, and of course Schumacher College, co-founded by Satish Kumar. Recently, Schumacher College has been going through a renaissance and a new Dartington Arts School has begun. It is wonderful to see how Schumacher College, for some years at the periphery of the work of the Trust, is now at the heart of all we are doing.

It is often the case that times of crisis
lead to creative responses and adaptation

In the winter of 2018, a group of people gathered in the beautiful surroundings of Dartington Hall in Devon. We were a relatively small number, with staff from Schumacher College and the former Dartington College of Arts. There were also leading thinkers and practitioners in the arts and pedagogy, including senior figures from MIT, the V&A, Carnegie Mellon University, The New School, and the journal *Leonardo*. We were there to try to identify where the leading edge of progressive learning in the arts is, and how we might once again manifest this at Dartington. In 1954, Hannah Arendt, who would become a professor at The New School, wrote an essay titled 'The Crisis in Education'. It was clear that the people gathered at Dartington in the winter of 2018 shared her view, that we are once again seeing a crisis in education. It can easily appear as if our current education systems are set up primarily to serve the economy – though I'm not sure they do that too well – and we can see from the climate crisis how statistics are unlikely to convince people to change habits and patterns. But there has arguably never been a more vital time to press for change. So, how and why might we effect such systems change?

In many ways the ideas explored and refined at the gathering in 2018 grew out of the progressive histories of Schumacher College and of the College of Arts. One of the last courses developed at the previous College of Arts was an MA in Arts and Ecology that started in 2003. This course prototyped pedagogic models and commitment to place, collaboration and whole systems learning. It came from a series of extended conversations with artists, scientists, economists, designers, engineers, and others who were all experts or leaders in their fields. Many said there were questions that interested them that went beyond the reach of their discipline. They talked about people from other fields whom they admired and who seemed to have insights that eluded them. We shaped the degree as a way to bring these people together in regular 'learning labs' designed to allow them to inhabit each other's practice.

The process was based on our collective desire to know the place deeply and to reach beyond our habits of seeing, thinking, and knowing, as well as on relentless curiosity and the wish to try to know the world through others. This encouraged connection and a love of place, and for kinds of learning and collaboration that incited a yearning for more. What became clear was that a close attention to place, any place, when approached through the prism of many ways of knowing, will reveal endless complexity.

Informed by these histories and dialogues we have redesigned all the learning programmes at the new Dartington Arts School and at Schumacher College. They are all focused on regenerative culture and based on the kinds of pedagogy I have described: short, intensive, place-based programmes with online and peer support, where collaboration and community, curiosity and compassion are critical to learning. Learning from and with each other is vital if we are to meet the challenges of our time. I would go further and suggest that love needs to be at the heart of learning. A love of the places we are in, with all their complexity and wonder, along with a love of learning.

A draft manifesto emerged from these recent dialogues:

Regenerative Learning at Dartington

- is committed to radical pedagogy ingrained in everyday life
- is necessarily a collaborative venture
- acknowledges that we are always in the field, the field is the place of learning, and we are the field
- is a network of learning communities grounded in ensemble methodologies
- actively creates space for uncertainty and ambiguity
- is learning to meet the other; challenging paradigms, preconceptions, disciplines, world views, and siloes
- informs, challenges, inspires organisational structures (and behaviours), including our own
- is immersed in the ecologies of place and thus engages with the great challenges of our time
- is an emergent choreography of integrative practices
- is immersive, embodied, experiential; 'felt on the pulses' (Keats)
- nurtures creative, compassionate, critical, playful, and contemplative practitioners
- incites a lifelong hunger and yearning for learning
- continually asks who is not in the room

The next phase of our exploration of regenerative learning is working to co-found a network of learning. We are placing an acre of land on the Dartington Estate into a commons arrangement, along with others in different bioregions around the globe, thereby connecting us with diverse projects and learning communities. We will enquire into deeply related and interwoven questions regarding soil health; biodiversity; water health and flood patterns; climate emergency; interspecies communication; human food futures; land rights issues; trauma, current and transgenerational; inequity and inequality; and much more that is likely to emerge.

We are not sure where this will lead but are committed to setting processes in motion that will support us in learning from and with a growing network committed to fundamentally reimagining our relationships and values, our place and participation in the complexity and wonder of our planet and the cosmos. Our hope is that these acres of land act as a place-based 'sourdough starter', to activate processes of change and regenerative learning that can contribute to the shift from crisis to global thriving.

Alan Boldon is Chief Executive of Dartington Trust and is Founder and Director of Weave, an international learning network created to nurture systemic wisdom; exploring creative processes with leaders from diverse backgrounds to unearth and share new ideas on how to build a better world.
https://weaveglobal.org

The School of Hope

Celebrating the Power of Youth to Transform Lives and Drive Grassroots Change

Gunter Pauli
with Ina Matijevic, Ivana Ražov and Martin Pavičić

*"Education is for improving the lives of others
and for leaving your community and world
better than you found it."*
– Marian Wright Edelman

Following the devastating Balkan Wars that started in Croatia in 1991 and then spilled over into Bosnia-Herzegovina a year later, the north Dalmatian village of Škabrnja, with less than 2,000 inhabitants located in the Zadar County of Mediterranean south Croatia, was left with a legacy of complete and utter destruction. It goes down in history as one of the darkest places in the five-year conflict.

The core of destruction in the village was the Vladimir Nazor School, where tragically, after the war, a mass grave was discovered in the school's grounds, along with a message painted on its walls saying: 'Welcome to the Village of the Death'. Such horrors engendered great emotional trauma not only in the town but also in the nation's collective psyche. The trauma was such that most of the survivors left their village homes. This impoverished and desolate village clung on with all the marks of war visible for any visitor to see. Memories of what happened, combined with high unemployment and landmines scattered everywhere, taking decades to clean up, set the stage for a grim future. That is, until some Croatian citizens decided upon a different future.

Decades after the war, there was still an undercurrent of aggression between Orthodox Christians and Catholics in this region. It was clear that this ancient feud between two Christian beliefs required a fresh start based on a novel approach to education, instead of repeatedly reminding everyone of the dramas and the victories of the past. So, a group of innovators gathered first in Geneva and then in Zagreb, the capital of Croatia, to discuss what would be needed to imagine and implement a transformation of society towards one that embraces peace. After consulting with Ina Matijevic, a food tech engineer, who specialises in systemic design and has passionately dedicated her career to unleashing transformation of human society through the wisdom of natural systems, a network of Croatian citizens committed their work towards providing a brighter future beyond anger and fear.

In March 2017, Ina Matijevic decided to approach the Skabrnja's School's Principal Mr. Marin Pavicič, who had personally escaped the massacre as a child. During the first meeting, Marin, the newly elected principal, shared his ideas and passion to do something extraordinary for the school and students. However, lacking in funds and being hindered by a political class that saw more benefit in maintaining the status of victimhood, little progress was made. Inspired by Marin's vision, Ina and her team set to work and sought help from many different sectors of society, sharing a vision for the school that was strongly supported by parents and students.

When I heard of this exemplary approach to the future of their school, based on the ambitious design and construction of the most ecological school in Europe, I encouraged the students to submit their concept to the Zayed Sustainability Prize, the United Arab Emirates' pioneering global award recognising sustainability and humanitarian solutions around the world. The formulation of their vision and their detailed application to the Zayed Sustainability Prize provided clarity to all concerned, and the confidence to see it through. Their ambitions rose and the team solicited additional funding from the European Union and Zadar County to implement their vision, which was dependent on an award yet to be won.

While the consultations and preparations continued to crystalise, the application marked the start of a transformational and an exciting journey that initiated the students' determination to embrace a 2,000-year-old tradition from the region. These 13- to 15-year-old teenagers founded 'Olive', a student-run organic olive garden cooperative. This area is renowned for growing excellent olives and the students' plan committed to strengthening this agricultural, cultural, and business activity. The students turned into entrepreneurs – entrepreneurs for the Common Good, including their own positive future.

The students turned into entrepreneurs
for the Common Good

On 15th January 2018, the Vladimir Nazor School was awarded the 2018 Zayed Sustainability Prize for the European schools. The Prize, which awards youth for their student-led projects with an emphasis on innovation, inspiration, and impact, recognised the students' desire to transform their school and their community with a commitment to install PV-T (photovoltaic and solar thermal systems) to provide electricity and water heating to both the school and surrounding community, taking advantage of the sunny climate of the area along the Adriatic Sea. The school committed to supply all excess power to the community, especially over the weekends and during holidays.

At the outset, the main technical objective was for the solar systems to generate 13,000kWh of electricity every year in addition to 60,000kWh of heat. The students also wished to cut their energy consumption by replacing all conventional light bulbs with LEDs. The 110 students from the middle school received solar backpacks, freeing them from carrying batteries to power their phones and computers when on a field trip, or walking back home from classes. An electric vehicle serves the cook at the school who prepares fresh food every day and is used for deliveries from the Olive cooperative. All students committed to study how to treat and recycle the school's waste.

Leonarda Škara, one of the students commented, "Winning this prize is the beginning of a new journey in my life, starting with our small school. The world of sustainability is our future and has opened our eyes and widened our horizons. It also gave us the discipline to perform and keep our promises. Our team participated in something so big that we could not imagine it. It turned into a true game changer." Leonarda hopes that their reversal of destiny will encourage others to challenge themselves because, as the students of Skabrnja experienced first-hand, a lot of great things can happen if one decides to just go for it!

The school utilised the $100,000 Prize to transform its grounds and facilities, which have now been completely renovated. The impact has been astounding. The original funds were complemented with additional money from the EU where the officials were enchanted by this resolute and positive approach against a backdrop of terrible trauma. With both the Emirati and EU Commission engaged, as well as local politicians who joined through the education secretariat of Zadar County, the students raised €750,000 an amount which at first was considered impossible to obtain.

This funded a full renovation of the school including the urgent removal of asbestos from the roof. Four years to the day since the school became involved in sustainable projects across multiple fronts, the school connected solar power on the roof to the grid on 9th of March 2021 resulting in the formation of a fully integrated system where the sun powering a 30kW of solar, providing electricity to LED lights, and LiFi internet using light instead of radio waves (WiFi) to transmit data.

This transition and modernisation positioned Vladimir Nazor as the first school in Croatia and one of the first in Europe to deploy the revolutionary LiFi to connect to the internet. It is also symbolic for connecting students through light in a school that is located only a few kilometers from the birthplace of Nicola Tesla, son of an Orthodox Catholic priest. This additional element in this portfolio of innovations gave renewed hope to the community – home to a 13th century Roman Catholic church – that there is a bright future. Some locals even went as far as to call this a resurrection.

The school's proactive steps go beyond green technologies. They put their vision of grassroots-driven sustainable development into practice by getting involved in other sustainable agriculture projects like cultivating mushrooms on the olive waste, with training provided by Ivanka Milenkovic, the Serbian mushroom entrepreneur, and feeding chickens with the spent substrate. These ongoing initiatives bring community engagement and human impact to the forefront of all their activities. This rebuilds community, social fabric, and economic resilience.

Recalling her personal experience in the Arabian desert when she visited Abu Dhabi to attend the Zayed Sustainability Prize awards ceremony in 2018, Ina Matijevic commented "This project was divinely guided from the outset. If you look at the world with your heart you will see amazing interconnections emerging. My wish is that Skabrnja speaks to heart of every student and teacher, in Croatia and any region where pain is felt. If Skabrnja can do it, anybody can do it".

The challenges on this road to transformation were many. The school (and the town's) tragic past, coupled with socio-economic difficulties and the more recent advent of the COVID-19 pandemic in Croatia, presented a host of obstacles that the school and its supporters had to overcome, but once their intention was set, momentum increased, and continues. Recently, the school has developed a connection hub for joint learning with 800 remote schools, serving students across the country. The school has equipped a new cafeteria for the students to eat healthy, organic food. The principal is committed to finding further funds to invest in a modern STEM (science, technology, engineering and mathematics) classroom for students, and an entrepreneurial STEM park with electric vehicle (EV) charging stations. And importantly, the site of the mass grave has been given a proper memorial stone, with candles lighting up the area day and night.

It is my belief that these students, teachers and this community have rewritten history. Just five years ago, the thought of anyone living in or visiting this area was unthinkable. Now we see in front of us a solar powered school and the largest organic olive garden in the country that will turn into the largest mushroom farm in the area growing one ton of mushrooms per day. This economic growth adding value to local and available resources is the real transformation.

It has become clear to all the participants in this project that this is only the beginning of a much wider transformation. The local municipality in Skabrnja has now decided to make 200 hectares of land available for a village cooperative, to build a solar power plant near the village to turn the whole community self-sufficient in power; and a local entrepreneur is planning to transform the old battlegrounds into the largest organic farm in the region.

What the school has offered more than anything else is hope – and it only has just begun!

 *Gunter Pauli is an 'Entrepreneur for the Common Good', and author of books on the transformation of the economy towards one capable of responding to everyone's needs with what is locally available. He has written 365 children's stories,* Gunter's Fables *which inspire the next generation to learn from Nature. He has implemented dozens of projects on four continents demonstrating that change is real.*

Flow with Life

Experiencing the Playful and Transformational Power of Nature as Teacher

Joseph Bharat Cornell

"Don't let schooling interfere with your education."
– Mark Twain

A teacher in the south-west USA once asked the children in his class to draw a picture of themselves. He recalled, "The American children completely covered the paper with a drawing of their bodies, but the Navajo students drew themselves differently. They made their bodies much smaller and included in the picture the nearby mountains, canyon walls, and dry desert washes. To the Navajo, the environment is as much a part of who they are as are their own arms and legs."

Nature's greatest lesson, I believe, is helping us feel our oneness with all Life. Nature cares for us in countless ways: she gives us sustenance, shelter, all our daily needs. Her most precious gift, however, is the experience of her harmony, wholeness, and joyful vitality. "In Nature," Gandhi writes, "there is a fundamental unity running through all the diversity we see about us." We cannot see this unity, invisible to human eyes, but we can feel it. Once, while camping in the Cascade Range, I entered a small subalpine valley covered with shallow streamlets and rafts of brightly coloured wildflowers. The loving presence in the valley was so palpable, so thrilling, that when it was time to leave, I had to drag myself away.

In my book, *Sharing Nature with Children*, I propose the teaching tenet: 'Experience First', to ensure that students connect with the world *as it actually is*, before attempting to analyse and interpret it. When absorbing Nature experiences, we feel a sense of timelessness and wholeness because our awareness is singularly focused and unified. Feeling calm is like a lake without ripples: it reflects life clearly. A calm feeling is intuition, enabling us to perceive life directly, as opposed to the intellect, which can only describe life. The word, 'ineffable' means 'beyond the power to describe' and profound Nature experiences are too subtle for the intellect to comprehend fully.

The right hemisphere of the human brain perceives the richness of the present moment and sees life in its whole context, and thus enables us to be fully aware of Nature. The left hemisphere steps away from the flow of direct experience (fed to it by the right brain) in order to analyse and re-present the world in an altered form. In this way, the left hemisphere offers clarity and functionality. In other words, the left's forte is to use and manipulate Nature; the right's forte is to revere it. Unfortunately, education today typically values left brain analysis, which reinforces the status quo and focuses on that which is already known.

As we grow beyond early childhood, it is essential that we stay immersed in the source of life itself. Just as snow in the higher mountains replenishes moist alpine meadows, enabling a luxurious growth of diverse flora, the right brain's heightened awareness inspires and vitalises the left-brain's interpretation of the world. Intuition's higher, more expansive awareness inspires all true works of art, all profound thought, and all genuine innovation.

As a young naturalist, I sensed the importance of play to break free of the limitations of the left brain and so I began to design play-centred Nature-awareness games. People would become so engrossed in the playful games that their experience would lead them seamlessly into immersion in the natural world.

An example of one such game comes from a recent workshop I held, where a participant, Johann, a professional German forester, described to me the profound change he felt in his relationship with the forest after participating in 'Sharing Nature' exercises. He told me, "My university training taught me to view trees as a commercial commodity. But now, after experiencing the 'Sharing Nature' forest exercises, I realise that the grasses are my friends, the trees are my friends, that every living thing in the woodland is my friend. This, for me, is a new way of looking at the forest. This awareness is going to fundamentally change the way I work."

During the Sharing Nature workshop, Johann interacted with trees in a variety of innovative ways. He and his co-participants, foresters from all over Germany, 'built a tree' together. Several foresters acted out each tree part—tap root, lateral roots, sapwood, cambium, phloem, and bark—and in doing so experienced kinesthetically the nature and function of that tree part.

Johann was then guided through a visualisation of himself as a deciduous tree, living through the seasons of the year. During the guided imagery, Johann planted himself firmly in the earth, spread out his branches, drew nourishment from the sun and sky, and turned air and light into life. With his sheltering branches, Johann cooled the summer air and warmed the winter air, thus making a more favourable environment for other life forms. Re-enacting a tree's existence and function enabled him to personally experience the role trees play in the forest ecosystem and to feel in himself many of the noble qualities of trees. By imagining himself living as a tree and nurturing the nearby plants and animals, Johann strengthened his sense of stewardship and love for the Earth. Adopting the role of a tree and offering sustenance to the life around him, Johann felt the energy of life flowing through his body and a marvellous sense of vitality, resilience, and wholeness. The variety of learning modes in this practice enhanced Johann's imagination, intuition, reason, empathy, and love, as well as his kinesthetic and sensory awareness, and thus enriched his appreciation and understanding of trees.

Touching people's hearts inspires their thoughts and behaviour. If our experience is mainly mental, our viewpoint on the subject tends to be materialistic. As a trained, practicing forester, Johann understood tree science well, but his scientific training had caused him (in his own words) to see trees simply as "a commodity." The intentional, multifaceted Sharing Nature exercise enriched Johann's whole being. As he experienced the forest in a more living, nuanced way, Johann himself became a more empathetic human being.

Over the years of sharing games as a teaching tool, I became aware that using the games in a particular sequence created a beautiful momentum toward a greater awareness of Nature. From that realisation came Flow Learning® — an outdoor teaching system that gently, almost magically, guides participants towards uplifting Nature experiences. Because Flow Learning uses play to create an outpouring of joy and enthusiasm, it creates wholehearted engagement with the learning experience and with the subject — Nature. This process makes leading a group deeply rewarding for the leader and for the students.

Flow Learning not only helps people experience Nature directly, but also helps participants develop their higher human qualities. Attunement with the natural world awakens the human heart. It is this inner development that is essential if society as a whole is to overcome its present challenges.

Attunement with the natural world
awakens the human heart

Flow Learning has four stages: *Awaken Enthusiasm*; *Focus Attention*; *Offer Direct Experience*; and *Share Inspiration*. These stages cumulatively awaken lively interest, heightened receptivity, and profound connection with the natural world.

The Flow Learning Sequence

I. Awaken Enthusiasm
Stage One lively and playful activities make learning fun, instructive, and experiential—and establish in the students an enthusiastic rapport with the teacher and subject.

Qualities: Playfulness and Alertness

- Builds on people's love of play
- Creates an atmosphere of enthusiasm
- A dynamic beginning gets everyone saying, "Yes, I like this!"
- Develops alertness and overcomes passivity
- Creates involvement
- Minimises discipline problems
- Develops rapport between participants, leader, and subject
- Fosters positive group bonding
- Provides direction and structure
- Prepares for later, more sensitive activities

II. Focus Attention
Stage Two activities, by offering simple challenges to the physical senses, help us become more calm, attentive, and receptive to Nature.

Quality: Receptivity

- Increases attention span and concentration
- Deepens awareness by focusing attention

- Positively channels the enthusiasm generated in Stage One
- Develops observational skills
- Calms the mind
- Develops receptivity for more sensitive Nature experiences

III. Offer Direct Experience

Stage Three activities create a deep connection with a natural site or object. These activities are generally quiet and profoundly meaningful.

Quality: Communion with Nature

- Fosters deeper learning and intuitive understanding
- Inspires wonder, empathy, and love
- Promotes personal revelation and artistic inspiration
- Awakens an enduring connection with some part of Nature
- Conveys a sense of wholeness and harmony

IV. Share Inspiration

Stage Four activities use the creative arts to clarify and strengthen personal experience. These activities nurture an uplifting atmosphere conducive to embracing noble ideals.

Qualities: Clarity and Idealism

- Clarifies and strengthens personal experience
- Increases learning
- Builds on uplifted mood
- Promotes positive peer reinforcement
- Fosters group bonding
- Encourages idealism and altruistic behaviour
- Provides feedback for the leader

Flow Learning's four stages lift players into a high level of aliveness, engagement, kindness, and kinship with fellow participants. Through 'Stillness-in-Nature' experiences, players find their senses and perceptions sharpened, their minds calmed, their receptivity deepened, and their understanding enriched by an awakened intuitive wisdom.

During a winter Sharing Nature workshop on Japan's northernmost island of Hokkaido, an eight-year-old boy sat quietly, intently writing a poem while all around him thick snowflakes showered down from the sky. So focused was the boy that he was oblivious to the cold and to the snow piling up around him. The snow rose to cover his legs, then his waist, and still the boy remained virtually motionless. Everyone else had long since departed for the comfort of a large, heated tent. Knowing the group was waiting for us, I asked the boy if he had finished his poem. "Not yet," he replied, immersing himself once again in his writing.

Like the boy in the snow, participants in Flow Learning delight in their heightened focus, calmness, and rapport with the natural world. The boy's experience exemplifies the psychological concept of 'intrinsic motivation' – motivation that is driven by rewards arising from within the person. Extrinsic motivation, on the other hand, is driven by external rewards which can include, as examples, the desire to earn money or to receive a high test-score.

Researchers in standard educational practices, though they prize intrinsic motivation, have concluded that, unfortunately, it cannot be easily summoned or sustained. Flow Learning, in marked contrast, has such a transformative effect on its participants that they *do* exhibit a remarkably high level of sustained interest and engagement.

A 2018 Flow Learning study involving 112 Taiwanese college students reported many beneficial results, which include:

- 96 percent of participants experienced an increase in their ability to feel Nature deeply
- 97 percent of participants were strongly inspired to love and protect Nature
- 100 percent of participants were 'able to open themselves to Nature with innocence.'

Summarising what Flow Learning participants experience, the author of the study, Dr. Hsiao Shu-Bih, poetically writes, "The heart is quiet enough to see what is most beautiful; the heart is quiet enough to hear the most beautiful voices. The heart is quiet enough to discover the original wisdom of life." Every human heart has the potential to be deeply touched by Nature. Flow Learning's metamorphosing process can help you as you share with others the joy and magnificence of Nature.

Flow learning can be used successfully in sessions lasting from thirty minutes to all day. Although it was originally developed for teaching outdoor Nature classes, it can be used to teach any subject matter, indoors or outdoors. It's four stages flow naturally from one to the next, and students respond readily to this sequence because it is in harmony with deeper aspects of human nature.

Each Flow Learning stage has its own characteristic activities: *Sound Map*, for example, is a Focus Attention stage activity, because players intently listen to (and mark on their map) the natural sounds around them. Participants — through joyful play and profound Nature experiences — typically become 100% committed to the class topic, a response which makes learning extremely rewarding for the students, and for the instructor. Such direct experience of Nature; intuition; creative, clear, and inspired thinking; intrinsic motivation; inner stillness; present-moment awareness; flow; and fluid, rapid change—all work together to infuse Flow Learning with power. When these deeper ingredients come into play, Flow Learning becomes truly magical.

Alan Dyer, long-time Flow Learning trainer and observer of its use in Europe, writes, "Just four headings with clear notes, backed by appropriate activities have been enough to provide anyone taking others into the natural world with a powerful structure to open eyes, hearts, and minds. And I mean *anyone*—from the very young to the mature, from the sceptical to the expert, and from the hesitant parent to the experienced teacher."

 *Joseph Bharat Cornell is the author of* Flow Learning *and* Sharing Nature with Children. *He is the founder and president of Sharing Nature Worldwide, one of the planet's most widely respected Nature awareness programs. For more information, visit:* www.flow-learning.org

CREATIVE ADVENTURES

HANDS-ON, PRACTICAL EXPERIENCE ALLOWS EDUCATORS TO TEACH BY EXAMPLE

Alan Dyer

"Education is our greatest opportunity
to give an irrevocable gift to the next generation."
– Ernest Fletcher

"There are only three ways to teach a child," the Nobel Peace Prize winner Albert Schweitzer famously told his medical students when they had to do some teaching as part of their course. "The first is by example, the second is by example, the third is by example." Maybe I should stop writing here and enter this as the most powerful tool for all teachers setting out to ensure their students 'get the message' of global care, responsibility, and action – but of course, it is not quite so simple, because we need to ensure that we have great teachers – fully trained, inspired, inspiring, experienced and exemplars of good practice. The UK Teacher Training Agency used the slogan, 'No-one forgets a good teacher' in a series of high-profile recruitment advertisements. But this is where it begins to get more complex, because training and supporting good teachers needs governments to provide the systems, infrastructure, finance and crucially, the philosophy to facilitate such training. So immediately, any aspirations for educational theory and practice become mired in politics and party manifestoes. To paraphrase the TTA slogan: 'Can anyone remember a good Education Minister?' And were they a good example?

Inevitably, political agendas revolve around improving and measuring pupils' academic achievements, league tables and international comparisons – with the consequence that the more holistic considerations are lost and seldom debated. The creative, exciting, and hands-on experiences that all children, students and adult learners deserve, is lost in spread sheets, attainment targets and learning outcomes that are restricted by a narrow academic focus.

> Educators know how comprehensive and important experiential activities are, and how they fundamentally influence student's lives

Changes to the political paradigm are, shall we say, 'ongoing' – they are complex, long term and beyond this essay. However, we already have the means to quickly enhance the educational experience across the globe for future generations by including into the curriculum a broader focus and a deeper understanding of themselves, how the planet works, and how to live in harmony with it. There is little work to do on the basic tools because we have been developing and refining them for many years across all continents and cultures. Educators know how comprehensive and important experiential activities are, and how they fundamentally influence student's lives.

It goes without saying that teachers are good at developing the core academic and specialist subjects – it is the primary focus of any government school system and an absolute human right that builds a literate, numerate society. However, given the opportunity, teachers are also brilliant at using the arts, crafts, music, drama, sport, games, outdoor education, cookery, spiritual education, gardening, and more focussed tenets such as Ecoliteracy, to build awareness, knowledge, understanding and responsible action. And we know that children love to choose some of these as their consuming passions and so develop an interest that can last a lifetime and even build a career. Yet many governments are cutting these subjects from their curricula or offering them only as an 'extra' which must be paid for privately. If we are to influence 'education for people and planet', our greatest task is to convince the bureaucrats and politicians to give teachers and children the time and resources to be creative.

Many schools already do a great job, particularly in the pre-school, early years, and primary classes (3–11), where there is more flexibility, and these subjects can easily be included in daily activity. What's more, these younger children generally 'get it' and have high environmental awareness, an openness to the issues and a love of learning by doing. Most education settings will now have some wildlife friendly space and a garden area or raised beds where children can grow and care for plants. Hopefully, they can engage in cooking and eating the produce they help to grow and follow the wonders of composting and re-cycling too. They maybe have an opportunity to visit local wild areas and engage with Forest School activities and express their feelings through art, craft, poetry, music, drama, dance, woodwork or writing.

This is wonderful, and it explains why the majority of environmental education programmes, books, projects and activities are focussed on this age group. But what happens at the next level – secondary/high school? Nothing! Well, almost nothing. At 11+ what happens are hormones, peer groups, over-stuffed curricula, exams, and huge pressures for teenagers – all of which, parents and educators often struggle to support. Coupled with these pressures, the holistic subjects with opportunities for residential trips, experiences outside of chosen subject areas and being taught practical skills are side-lined and only available as paid-for extras for those who can afford them or even worse, they are abandoned. I passionately believe that giving children (and *all* learners of whatever age for that matter) the opportunity to be creative, to experience awe and wonder, to experiment with solitude and emotions, to grow and nurture beautiful and edible plants are absolute rights.

But again, of course it is not so simple. Children need the wisdom, guidance, advice and often the permission from their family elders, from teachers and the wider educational establishments ... and of course it all needs funding. How that guidance is given to children is another long essay, but the important word here is *opportunity*. We must engage on all levels to provide a more holistic and creative education within the curriculum and help students to experience these wonders on their own terms.

Of course, we must also keep students safe – they don't need to see the risk assessments or the safeguarding policies, but they (and their parents) need to know they are in place. I remember attending an environmental education conference and listening to a keynote lecture about safety given by the CEO of RoSPA - Royal Society for the Prevention of Accidents. He opened by saying, "We want all children to have accidents..." Stunned silence enveloped the room for a while, but then he explained that taking risks is an important part of growing up and the odd cut, fall or bruise is absolutely okay, but more serious risks-taking with the educational sphere can still be quantified, understood and controlled.

So, if we are to 'educate as if people and planet matter' then all of us in all sectors of society need to engage with educators and to take action on many levels. It is not something *they* do, good education is not just the responsibility of governments, ministers, teachers, parents, environmental or ecological educators – quite simply it is up to all of us. Teachers, lecturers, professors and group leaders will, however, need to check a few things first – what I call the 'Yes, Buts'. In over 50 years of running trips, adventures, expeditions, events and workshops, no-one has ever said, "No! That is not right". They *have* said, "Yes, but... have you filled in the... are you qualified... where is the funding... do you have permission...?" So, once you are first aid trained, have answered the above, have all the right risk assessments in place, then you are ready to forget the 'buts' and be that inspiring example.

If you need some starting points, how about:

- Taking your group out at night – not just camping overnight but getting close to the inhabitants of the dark and experiencing all of the amazing moments from dusk till dawn.
- Revisiting the same natural place every month, or at least every season, for a whole year. Each person can take time alone in a special spot and record what they see, hear, and feel in any appropriate way.
- Get a team together to make a real difference to a natural area – help dig or rescue a pond, plant trees, make nest boxes, restore a meadow, protect a riverbank... ask your local wildlife group; they will have a list of projects just waiting for you.
- Grow food – whatever is appropriate to your bioregion; understand it, cook it, eat it and share it each season.
- Get some tools and materials together (often given free if you ask), teach your group how to use them and watch them grow in confidence as they make, repair, and create.
- Celebrate Nature and futures with a festival to share the arts, crafts, dance, drama, music, writings, and progress you have created – invite the whole community.

Today, more than ever, there is a vast range of books, videos, ideas, and practical advice available to help you try new adventures which are truly transformative. We urgently need to immerse our learners in real, seriously hands on experience which has the capacity to help them engage with Nature, protect, conserve, and enhance the integrity of our precious planet Earth.

Then maybe we can all lead... by example, example, example.

 *Alan Dyer taught biology for almost fifty years, then became a lecturer in environmental education and Director of the Centre for Sustainable Futures at Plymouth University. Alan now runs Axewoods Co-operative – a volunteer organisation for woodland management, log banks and community learning.*

The Role of Happiness
in Environmental Education

LEARNING HOW TO PRACTISE
WITHOUT PREACHING

Isabel Losada

*"There is a moment where you have to choose
whether to be silent, or to stand up."*
– Malala Yousafzai

Life is short. Different spiritual paths put different
emphasis on why we are here. Christianity suggests the
reason for our existence is to learn about Love. Judaism
points to 'Tikvah' (Hope). And Buddhism? Well, I
used to do a talk in which I told the audience that His
Holiness the XIV Dalai Lama, when asked the purpose
of life, summarised the complex teachings of Tibetan
Buddhism into one word. With a large grin, I ask the
audience to shout out what they believe the word might
be. The answers are often surprising. It's true that some
will shout out 'Compassion' or 'Life itself' while some
shout, mysteriously, 'the breath' or 'transformation.'
Some will even say 'death.' What do you think? The
answer - given by The Dalai Lama, is 'Happiness.' As he

is considered to be one of the greatest Buddhist teachers of his generation, I think we can assume he has given the matter some consideration.

But anyone who has been studying or teaching environmentalism in any form for the last five years could be forgiven for thinking that the answer is 'Suffering.' I'm not saying that the problem isn't very serious, but please don't let's teach the history of the problem, the scientific measurement of the problem, the economic entrenchment of the problem, the international complexity of the problem or the likely impossibility of solving the problem. It doesn't help. It doesn't help me anyway. It's not our job as educators to discourage and inspire climate anxiety and despair. I don't want to learn about the problem. I have an obsessive single-minded focus – the solution. Please – teach that.

In a recent book, *The Joyful Environmentalist: How to Practice Without Preaching* I educated and entertained myself, (and hopefully the reader too) with a complete life overhaul. And this 'out of the classroom' journey has to be enriching because, as I mentioned, life is short. So, I looked at every single aspect of how we live, firstly to explore how we can look after the planet and secondly how we can do that in a way that makes our lives more real, more sensual, perfumed, rewarding and inspiring. How we can make it better: in every way. In my experience these two quests go together well.

Let me offer you a few examples. For instance, ethics and the way we bank. I've been with the same bank since I was a teenager. I've always had a nasty feeling that they are probably investing in fossil fuels, funding local vivisection facilities, giving money to the arms industry and all manner of dodgy dealings not in line with my values. So, I wrote and asked if their lending and investment portfolio is in the public domain. The short answer is no. I consulted *The Ethical Consumer* for the most ethical bank and found that they recommend Triodos bank. I wrote to Triodos to ask if their lending and investment portfolio is in the public domain? "Yes," they said and showed me where to look on their website. There were wind farms, permaculture projects, small organic agriculture initiatives – all businesses I'd be happy to support. I changed banks. "But surely there is nothing joyful about moving your finances?" asked a sceptical radio journalist. Oh, but there is – for the first time in my life I have a bank I'm proud of and I'm 100% sure that I'm not lending money to the fossil fuel industry. I never learnt about my power as a consumer when I was a student. Yet empowering young people with knowledge of the differences that they can make is one of the most enjoyable aspects of an educator's role.

Slowly but surely, I removed all plastic from my life – and how beautifully my home is transformed. I have hand-woven baskets from charity shops instead of nasty mass-produced plastic boxes, UK-grown seasonal vegetables arrive at my home in cardboard boxes that can be returned. It's possible to make a game of becoming waste free. I don't have a car. In researching this I discovered how cars have destroyed our communities as well as the atmosphere. If more people didn't have cars, they would choose to live closer to the people they love. Finding alternative ways to travel is one of the games I enjoy most... but we need to be radical, otherwise (as they say at Friends of the Earth) we are going to hell in a hybrid. Can we teach that cargo bikes are more fun than cars and you don't have to take out a gym membership?

> Empowering young people with knowledge of the differences that they can make is one of the most enjoyable aspects of an educator's role

So many people seem to think that living more environmentally involves purely sacrifice. I've found the opposite. We environmentalists are arguing for a better quality of life and ultimately for the abundance of life itself. We can discover our passions and combine those with the kind of environmentalist we can become. Please, don't let us bore young people into passivity by telling them what they must not do and can't have.

If we want to 'Practise Without Preaching' then we need to put environmentalism first in our own lives until the way that we live is so radically different that we inspire others. You know the old 'be the change you wish to see' solution – it's the only way. And it has to be a joyful way. One day while enjoying a little activism (by playing the drum in a samba band to support a local environmental group lobbying for better air quality) a young woman expressed a desire to support the Earth but said that she didn't know how. "I'm just a gardener," she said. I chatted to her about re-wilding, biodiversity, the 'Insectageddon', and the vital role that gardeners can enjoy.

Finding what young people (and their parents) are passionate about is simply a question of watching their energy. Some people only come alive when they talk about cooking. So, what is the path of the environmental chef? Many who understand animal agriculture have adopted a vegan or plant-based diet. When posting about a delicious vegan recipe on Instagram yesterday someone wrote underneath, 'Your mind, body and soul feel so much better being vegan.' And that's also what we want to teach isn't it? Creating minds, bodies and souls that feel better? "That's easier in an educational context, but how do I deal with my husband/wife/mother /son?" I'm often asked. I always reply, "by delighting their senses". And by that, I mean with food – seduce people into eating for the environment. Prepare such amazing mouth-wateringly delicious vegan food (curries with quality spices are my personal favourites) that you put the focus on all the wonders of the organic, seasonal eating world.

Education and clothes? Clothes have become so damn boring. To quote one friend, "Everyone looks the same." And we already have enough clothes. All of us. How do we know? Well, we don't see many people walking the streets naked, do we? So, if we are going to buy clothes then let's let fun be our guide. And creativity. And of course, environmental awareness. There was a time in the 1960s when people took delight in self-expression through what they wore. People enjoyed colour and fabrics. Now black is the new black is the new black. Today environmentally aware, creative, savvy young people go to charity shops and play with sewing machines and such creativity has long been known to link directly with positive mental health. Or can we even imagine clothes made from organic cottons, dyed with natural dye that don't hurt our rivers, that we've made ourselves or someone has made specially for us – clothes that contain that magic ingredient: love. If we are going to buy new clothes, then exploring supply chains for our own education will help us to find clothes that we can feel proud to wear. Can we teach this? Can we teach this by example?

And then of course there is Nature. I saw a butterfly in my local park today. I don't wish to seem ungrateful but it's the only one I've seen this year. This park isn't natural – it's cut back, trimmed leaf blown and filled with weed killer... if I was a bug there would be no-where to run to. And the birds in the park. Well, we have crows, parakeets and some pigeons. This is barely an imitation of what we could enjoy. Free Nature, where people have had the guts to leave everything to grow as it pleases is a magnified experience. I've experienced rewilded Nature at the Knepp Estate in West Sussex, I've seen beavers swimming in the rivers in Devon. Being in this kind of Nature – real Nature where humans are not in charge, makes you glad to be alive. Just to celebrate my point, in the dawn chorus at Knepp at certain times of the year, you can hear yellowhammers, Cetti's warblers, whitethroats, swallows, swifts, great tits, blue tits, cuckoos, spotted flycatchers, crows, magpies, jays, linnets, red kites, treecreepers, green woodpeckers, peregrine falcons, robins, blackbirds, buzzards, kestrels and even storks. Moving towards a richer Nature heals our soul profoundly. When we visited Knepp, my daughter said, "Here everything feels right with the world." Part of our job as educators is surely to ensure that young people experience this? As has been said many times, "Why would anyone strive to save Nature if they don't first love Nature passionately?"

And so, as we move forward as environmentalists, let's let happiness guide us. I'm one of those who believes that no action is too small and no project too large or too ambitious. On the same day I'm buying a bamboo toothbrush as a gift, I'm also endeavouring to find a way to get my entire street (18 houses) to agree to having solar panels fitted and lobbying the local council about CO_2. When we work with young people, we need to ask them each what they most love to do and find a way to support them in developing that passion whilst being engaged environmentalists. But please, no ideas that make them feel bored – no ideas that make any of us feel bored. We need to be radical, we need to be in a hurry. We need to be environmentalists and we need to be happy. Because it's a wonderful world. And life is short.

Isabel Losada is author of seven books including The Joyful Environmentalist *and* For Tibet with Love.

A Universal Curriculum

AN EDUCATION FOR EVERYONE,
FOR ALL TIME

Colin Tudge

"Not everything that is faced can be changed;
but nothing can be changed until it is faced."
– James Baldwin

Whoever would presume to influence the minds of others must tread fine lines between conflicting needs and ideals. Every curriculum must have a point of view or else be simply inchoate, yet true education must not be doctrinaire. It should introduce the widest manageable range of thoughts and possibilities while also equipping students to critique whatever they are told and to decide for themselves what to take seriously.

We must seek too to reconcile the contrasting demands of true education, which implies a 'leading out' into a broader understanding; and training, which requires would-be learners to focus on particular and prescribed skills and ideas. Tennis players and musicians, however educated, need training. But I have also heard scientists and even philosophers claim to be 'trained' and although both must learn some special techniques, they surely must above all be educated, leaders of thought – for if not them, then who?

Yet the distinction has often been taken too far and separated the putative thinkers from the doers – inevitably leaving the power with the former, which exacerbates the inequalities of class and wealth that, together with racial and gender prejudice, are the prime sources of injustice and strife. As John Ruskin commented in *The Nature of Gothic*: "We want one man to be always thinking, and another to be always working, and we call one a gentleman, and the other an operative; whereas the workman ought often to be thinking, and the thinker often to be working, and both should be gentlemen, in the best sense."

Education too should be relevant to life as it is experienced at any one time and place – yet whoever knows only the world around them is doomed at best to be a follower of fashion. We need to be able to step outside the zeitgeist for, as Rudyard Kipling asked, rhetorically, "And what can they know of England, who only England know?"

Finally, traditional education has tended to reinforce and largely to create divides between the various academic disciplines – notably between science and the arts, and between science and religion. Yet the relationship between them all should be and often has been synergistic. Indeed, all disciplines must be pursued in the light of all the others to provide a truly 'holistic' understanding.

With all this in mind, some of us are seeking to devise a new kind of curriculum and indeed to found a new college offering a different kind of education. The curriculum is divided conceptually into four tiers: GOAL, ACTION, INFRASTRUCTURE, and MINDSET. Each tier is then further divided into subjects – twelve in all. Each is then discussed from first principles – meaning the 'bedrock principles' of morality, which asks, 'what is it right to do?'; and ecology, which asks, 'what is it necessary to do (in order to do good) and what is it possible to do, within the limits of planet Earth?' Indeed, all human action should be guided by morality and ecology – both rooted in the ideas of metaphysics.

We suggest – and this is the nearest we come to doctrine – that the GOAL of all humanity must be to create convivial societies, with personal fulfilment within a flourishing biosphere. In short, society matters, individuals matter, and the biosphere matters. Present-day governments rarely spell out what they are trying to achieve, and why. They regale us instead with chauvinistic slogans, aimed at wealth and dominance. But our stated goal, like everything else we say, is up for discussion.

ACTION includes all technologies but the most important by far is that of agriculture. Farming is at the heart of all the world's affairs, from social justice to global warming and mass extinction. The fate of humanity and of our fellow creatures depends on how we farm. But agriculture, like everything else, is nowadays dominated by an oligarchy of big governments and transnational corporates and their selected intellectual advisers and, in neoliberal vein, the oligarchy sees agriculture just as another, not very efficient way of making money. If we seriously care about the long-term future, we need to develop Enlightened Agriculture, which combines the ideas of agroecology (treat all farms as ecosystems) with those of food sovereignty (every society should keep control of its own food supply).

Both lines of thinking – agroecology and food sovereignty – lead us to favour small, mixed (polycultural), low-input (basically organic) farms that perforce are complex and so must be skills-intensive (plenty of farmers and growers) and so in general should be small to medium-sized. This is the precise opposite of the high-input, ultra-simplified monocultures with minimum to zero labour that are now promoted by the oligarchy. Alas, much or most of what is now advocated from on high in all contexts is the opposite of what humanity and our fellow creatures really need. Agriculture, too, more than any other discipline, combines thinking with hands-on involvement, as Ruskin advocated. As Adam Smith commented in *The Wealth of Nations*, in the century before Ruskin, "… no trade requires so great a variety of knowledge and experience."

> Alas, much or most of what is now
> advocated from on high in all contexts
> is the opposite of what humanity
> and our fellow creatures really need

Agriculture needs a corresponding food culture if it is to thrive, and here we encounter huge serendipities. For enlightened farms must focus first on arable and horticulture, fitting animals into the ecological gaps, and since such farms are also designed to be diverse, they produce, overall, 'plenty of plants, not much meat, and maximum variety'. These nine words summarise modern nutritional theory (modest protein, high fibre, low fat) and also the basic structure of all the world's finest cuisines as practiced on an axis between Italy and China, which all use meat sparingly. So, we don't need to be vegan, and we don't need lab-grown ersatz. We just need good cooking – which should be given high priority.

INFRASTRUCTURE includes governance, economics, and the law. All must be geared to the grand goal of conviviality, fulfilment, and the wellbeing of the natural world, guided by the bedrock principles of morality and ecology. Again, at present, this is far from the case. Above all, economics must be rooted in morality, as Adam Smith and Keynes and Marx were well aware. The neoliberals, who are currently dominant, claim to be 'morally neutral' which is both pernicious and absurd.

Finally, I suggest, MINDSET – the sum of all our preconceptions and attitudes – has four prime components. The first is morality – which is not simply 'relative' (dependent on cultural norms) as it has been fashionable to argue. All the transnational and most of the Indigenous religions that survive agree that the prime moral principles – the prime virtues – are those of compassion, humility, and an attitude of reverence towards the natural world. These, therefore, can reasonably be taken as the basis of a universal morality.

Science, the second prime contributor to our mindset, should be taught not simply as the source of 'high', technologies but as a cultural – aesthetic and spiritual – pursuit. The philosophy of science must be taught alongside to remind us that despite the wonders of its ideas and achievements, science has severe limitations and is not the royal road to truth. Yet truly modern science has given us some very positive insights of a metaphysical kind. Thus, modern biology tells us that Nature is more cooperative than competitive – superseding the crude, brutalist interpretation of Darwin which, alas, still prevails. Modern physics has given us the idea of universal consciousness – that consciousness is not created *de novo* in the human brain but is a quality of the universe of which we partake.

Metaphysics has been all-but abandoned as an independent discipline, yet it addresses what are often called 'the ultimate questions': What is the universe really like? How do we know what's true? What are the roots of goodness? Metaphysics introduces the essential ideas of transcendence and the sense of the sacred; of intuition and hence of mysticism – by-passing the intellect and tuning in directly to the universal consciousness. This, surely, is the true meaning of spirituality. These notions are at the core of all religions – and are shared by all of them.

Finally, the arts are the human mind in free flight, conceiving and making apparent the otherwise unimaginable. They are the essential joker in the pack.

Our curriculum is infinitely flexible. It can be introduced in outline in an hour or broken into individual topics to be permutated and re-combined at will. The ideas may be pitched at any level from kindergarten to post-doctoral, for professionals or people at large, young and old. Teaching primarily is by dialogue, with everyone brought into the conversation. Our new College could have its own premises or could, as now, be on-line and peripatetic, contributing to established courses in established centres of learning.

I suggest this broad, holistic, dialectic, open-ended, flexible education that is relevant and available to all is vital if we are to heal the rifts that divide humanity and have so disastrously separated humankind from the rest of Nature.

Colin Tudge is a biologist by education and a writer by trade with a particular interest in food and farming. Ten years ago, he co-founded the Oxford Real Farming Conference and from that has emerged the College for Real Farming and Food Culture. www.collegeforrealfarming.org

Lessons from the Periphery

Embracing non-linearity, non-locality, unpredictability, and active co-organisation, where 'learning by doing' is at the beating heart of the educational experience

Pavel Cenkl

July Mountain

We live in a constellation
Of patches and of pitches,
Not in a single world,
In things said well in music,
On the piano, and in speech,
As in a page of poetry –
Thinkers without final thoughts
In an always incipient cosmos,
The way, when we climb a mountain,
Vermont throws itself together.
– Wallace Stevens

In more than 30 years of running across mountain ranges in North America and Europe, I have seen the sunrise from more summits than I can remember — from snowy

253

peaks in the US northeast, to the re-emergence of the not-quite midnight sun in the Arctic, to the sun piercing a misty moorland from tor-capped hills in southwestern England. The brilliance of that singular moment when the sun sets the world aflame in swathes of goldenrod, amaranth and rose and gives shape to the land and brings the world into relief. As much as each of those moments remains indelibly etched in memory, they are also always painted on a canvas without end — one which inevitably includes rising in the chill of the pre-dawn, often running up through a rocky, icy, or bog-filled slope by torchlight; the familiar company of silence, breath and wind; the movement and effort and breath and sweat and sliding of muscle under skin; the feel of changing ground underfoot; the body as it generates heat on the journey upwards. Each of these discrete moments, actions, and decisions compose for me the whole of the experience, and finally, invest the sunrise with meaning broader and deeper than it might have otherwise had. As the world throws itself together in this light of a new day, the landscape comes into sharper focus, and the moment itself becomes a layered palimpsest of experiences — an instance of "here" that is always "shot through with there," as Timothy Morton writes, where "our sense of place includes a sense of difference."

As Head of Schumacher College and the Director of Learning at Dartington, on any given day, my job consists of many things, and as I reflect on what I learn in some of my most cherished moments, when I am most connected to the work of developing programmes, designing courses, and outlining the teaching sessions

I am privileged to share with students, I find that this image of the mountaintop sunrise, this moment of literal enlightenment, offers ground just solid enough from which to look out at and begin to see a world where the edges between the experiences of past and future, centre and periphery, self and world imbricate. Like the moment of looking out from a summit onto a world newly illuminated, we can no longer afford to keep the experience of learning at the periphery of education. It must take its place at the very centre, in a model wherein the oft-touted 'learning by doing' is not relegated to the margins as an ancillary activity that's extra or co-curricular. The role of experience is not merely to supplement the information exchange typically at the core of the higher education curriculum, but is itself at the centre of a learning paradigm where meaning-making is embedded in the fabric of relationships, practices, and interactions. Put simply, authentic meaning-making through relational experience is the central element of any meaningful learning.

> We can no longer afford to keep the experience of learning at the periphery of education.
> It must take its place at the very centre, in a model where 'learning by doing' is not relegated to the margins as an ancillary activity

Engaging with the more-than-human world is a foundational practice of the learning community at Dartington and Schumacher College and is embedded in every postgraduate and undergraduate programme as well as in the rhythms of daily life on the Estate. As we look at learning more broadly across higher and further education, can we leverage our focus on ecological design and look to the ecosystem as a model to help us rethink and reshape education? Ecosystems are resilient, adaptive, open, always evolving, and are distributed across places, individuals, and species. No one thing is at the centre of an ecosystem, but rather, relationships, processes, and networks. Often, surprising generative interactions are its essential building blocks. Why then should there be a centre to a system of learning? Rather, a system of learning that is networked and distributed that is modelled on ecological systems could foster within a learning community active and co-creative engagement with delivery, projects, assessments, and co-designing students' very course of study. By subverting the centre of learning, the curriculum could – and should – embrace non-linearity, non-locality, unpredictability, and active co-organisation.

Developing learning programmes and curriculum is always a process of weaving together relationships, collaborations and exchanges, an embrace of complexities and a letting go of certainties in a process of co-creation. Designing a curriculum is processual — an incomplete and always evolving engagement of ideas and practices that draws us out of ourselves in ever-widening gyres along an experiential arc that brings us back again, enriched, empowered, and engaged. The ecologist and philosopher Timothy Morton describes hyperobjects (in his book of the same name) as so indescribably massive that they defy definition, although they are everywhere and touch our lives at nearly every turn. We can only know these complex hyperobjects often through the smallest of tangible manifestations — for example, unseasonably heavy rains as the embodiment of climate change; the takeaway coffee cup as our connection to global commerce, overburdened waste streams, colonial histories, and centuries of unequal economic relationships; and the mobile phone a manifestation of resource extraction, labour inequities, and changing social norms. The hyperobject is something that continually intersects and interweaves with human and more-than-human lives but surfaces only in moments of interaction. Education must encourage the experience of and response to that arc of interaction.

Since early 2020 and the beginnings of Covid-19, we have seen a dramatic restructuring of learning in response to the global pandemic, and while at least initially this effort was necessarily both reactionary and responsive in order for institutions to continue delivering courses, it has also provoked serious debate about the very heart of higher education. Many educators, writers, and educational theorists have raised questions about whether the existing university model is sustainable, whether learning should move largely online and away from in-person teaching for the long-term, whether international student travel and consequently in-person enrolment is either financially or environmentally sustainable or ethical, and, fundamentally, whether higher education as we have known it for centuries will or should survive this crisis.

If we are able, this is a moment in which to pause and reflect on how we've arrived here and where, if we have a choice, we would like to go next. Paul Friga, Associate Professor at UNC writes, "This is no time to be incremental or reactionary … this is the opportunity of a lifetime for enacting positive change in higher education." Nearly all universities have been challenged with balancing short-term crisis response with longer-term thoughtful planning. Initially, 'everything online' is a reaction necessitated by global school closures and institutional desire to finish the academic year and support students as they complete modules, courses, and programmes. The prolific writer and speaker on educational institutions and collaboration, Clay Shirky, has said, "the distinction I've constantly tried to draw when talking to deans and faculty is between remote instruction versus courses designed to be delivered online." The former is reactionary; the latter, if done well, revolutionary.

To place our current crisis in an even clearer context, Shirky wrote in 2014 that our biggest threat in higher education is not, "video lectures or online tests. It's the fact that we live in institutions perfectly adapted to an environment that no longer exists." What higher education's uptake of technology has enabled is a reorganising of centre and periphery by involving more voices and diverse perspectives and student experiences to share in rich practice-based learning across cultures, continents, and ecosystems.

As many countries begin to re-emerge from pandemic restrictions on travel, learning, and the ability to gather, we are at a pivotal moment in higher education: we have an opportunity to be decisive and shape a future that is less dependent upon historic forms and structures of how learning is delivered and is more creative, flexible, accessible, engaging of real-world ecological and social challenges, and grounded in student experience.

A key part of our approach must be to rethink our basic models of teaching and learning so that they are far better aligned to support flexible schedules, varying abilities to pay higher and higher tuitions, and balance demands of immersive in-person residential study with self-paced learning at home. In short, we need to be able to build vibrant and thriving learning communities that present a wide range of learning experiences and options to help students achieve their goals and ask difficult questions, including:

- Which elements of the way we design, develop, and deliver higher education are essential?
- Which have been historically flawed and have been ignored for far too long already?
- To which do we return only at our peril?
- And, ultimately, what could higher education look like if we understand and engage with our world as a complex, integrated socio-ecological system?

Higher education is acutely in need of a framework that enables and empowers the integrated use of existing tools and approaches to reach a deeper understanding and engagement of meaning-making across cultures, pedagogies, worldviews, species, and ecosystems. As one form of response, a multi-scale learning network is always already in the process of co-becoming, manifesting a world in which organisms communicate always in an unfinished processual dynamic. Our covid-influenced present and unpredictable future demand radical revision of higher education's traditional forms of delivery. An ecosystem-scale approach to scaffolding distributed site-based learning can help make a pathway toward a resilient, adaptive, and multi-scale curriculum.

A regenerative approach that continually enfolds, adapts, and participates in complex socio-ecological system dynamics through acts of interspecies listening, co-creation, and collaboration can help build a more resilient and relevant model for higher education around the world. It can enable us to recognise that, whatever our discipline or specialism, we are always teaching about relationships — between human and more-than-human, among people in community, about the relationships among things rather than just about the things themselves. We live in a constellation 'Of patches and of pitches, Not in a single world' as Wallace Stevens writes, where even as a brilliant sunrise sharpens the edges of the subtle softness of a pre-dawn world, what comes into relief are the emergent, vibrant patches of aliveness – the relationships at the periphery that are the essence of our experience.

As colleges and universities tentatively envisage a return to a 'new normal' of learning in a post-covid world, we cannot afford to squander this opportunity to rethink our basic models of teaching and learning. We already have the building blocks to build vibrant, thriving, flexible, and co-created learning communities that can help us all to leverage a wide range of learning experiences and options to deliver education as if people and planet matter.

Dr Pavel Cenkl is Head of Schumacher College and Director of Learning at Dartington Trust in Devon, England. He has written and presented widely in the areas of curriculum design, pedagogy, environmental humanities, and ecology. He is an avid endurance athlete, and his Climate Run project draws attention to the intersection of movement, ecology, and climate change. His books include: This Vast Book of Nature: Writing the Landscape of New Hampshire's White Mountains; Nature and Culture in the Northern Forest; Transformative Learning: Reflections on 30 Years of Head, Heart, and Hands at Schumacher College *(with Satish Kumar).*

The Light of Learning

GREENING THE SCHOOLS OF BHUTAN

Thakur S. Powdyel

*"The highest education is that which does not
merely give us information but makes
our life in harmony with all existence."*
– Rabindranath Tagore

In the beginning was the promise. "Come," we told
the child, "we will build a school for you. We will
bring the light of learning to you, and the welcoming
doors of exciting discoveries will open wide for you.
Come, let's make meaning together and weave our
dreams in common."

It has been a long journey and we have been on the
road for millennia. Somewhere along the way, the rains
started beating upon us and we lost the child and forgot
the promise. The vision was gone, and we all became
children 'lost in a world-fair.' We got caught up in an
impoverished role and stood culpable witnesses to the
slow extinction of the nourished soul.

All is not lost, though. Life remains, the Planet calls,
and the future beckons. Let's rebuild the ship, collect our
oars and set sail again. We are here not to lament the loss

of paradise but to share hope and to build faith. With all its imperfections, the world is still a good place and "has enough for everybody's need," as Gandhiji assured, and despite all its pains and privations, life is still beautiful and precious and every bit worth preserving and worth celebrating. And the Earth yearns for our loving embrace. We owe it to our children and their children and beyond to speak from the core of our being and to serve them with the integrity of our thought, speech, and action. We owe nothing less to our Planet Earth and to Life. That's why Education matters.

We must follow the light as the wise ones have shown and reclaim education as a gentle process to guide the mind to look for and to love what is true and good and beautiful, even as we look ahead and engage the limitless possibilities of science and technology.

We have some mending to do here, to repair broken relationships, and sort out lopsided priorities before we move on. For all the progress the human race has made on all fronts over the years, we find ourselves in a visionless world, plagued with the unconscionable inequities of an unmitigated rat-race, with the consequent degradation of the natural world creating conditions that are incompatible with a meaningful life, against the backdrop of an ailing planet.

We may begin by restoring the integrity of our beloved Planet Earth that is our home, which we share with an infinite variety of other beings and from which we draw our vital life-force and succour. A supreme irony of our times is the severance of the precious relationship between humans and our all-giving Mother Nature, resulting from the arrogance and exploitative impulse of our so-called most highly evolved of the species, that sees this Earth merely as a resource to feed the utilitarian, reductionistic GDP-machine, which has come to be viewed by many as 'gross distortion of progress'.

Healing the gaping wounds of this fractured relationship calls for a saving grace of humility and an attitude of reverence towards the infinite bounty of Planet Earth against the finite capacities of humans. If we align our little heartbeats with the giant pulsations of Mother Nature, we may indeed achieve restoration and heal the broken bonds that sap vitality and threaten life.

Many of our hitherto sustaining foundations are giving way, but we still have the promise of education to help in the great healing. Obviously, it must be education not as it has been, but as it ought to be. Today's educational experience remains largely scattered and fails to discover the inherent harmony underneath all phenomena. Given the disillusion and anxiety that young and old alike face, our seats of learning will do well to return to the nourishing care of Mother Nature and discover with Blake:

"...a World in a Grain of Sand
And Heaven in a Wild Flower,
Hold Infinity in the palm of your hand
And Eternity in an hour..."

This act of making meaning, this movement back and forth between the particular and the universal and vice-versa is a most powerful achievement in learning. In much the same way, the learner can be guided to find with Shakespeare:

" … tongues in trees, books in the running brooks
Sermons in stones, and good in everything"…

And indeed exclaim with Carl Sagan: "The Earth: that dot! That's here! That's home. That's us."

This is restorative learning, deep, meaningful learning. All disciplines have in their essential nature an all-harmonising principle that can lead the seeker to the Promised Land but is routinely lost in the scramble to complete the syllabus and move on. It is unfair to the giver and the receiver. Education must aim higher and be more sublime; it must lift the mind, expand the heart, and instil faith where cynicism spawns.

Mother Nature presents the most detailed curriculum, perfect in every sense, relevant to all time and all space, replete with never-ending magic and ever-renewing miracles, integral, self-sustaining, all-purpose. She is the university *par excellence* and the teacher *non-pareil*. At a time like now when restless technology has replaced the slow yet insightful process of making meaning with the instantaneity of the search-engine, education can regain its grace and relevance by holding Mother Nature as its abiding point of reference.

> Mother Nature presents the most detailed curriculum, perfect in every sense, relevant to all time and all space, replete with never-ending magic and ever-renewing miracles, integral, self-sustaining, all-purpose.

It is a sad commentary on our times that the world sees people largely as an economic resource whose value resides in their 'doing', not in their 'being'. Our young men and women are defined by dehumanising market-metaphors that ask: Are they marketable? Fit for job? Or worse still: Are they saleable? And our temples of learning become production-centres to feed the soulless greed of factories and corporations. But people matter. They animate our world, our nations, our societies, institutions and homes, from continent to continent, coast to coast, on all parallels and meridians. Our saints and seers, scientists and philosophers, inventors and discoverers, artists and entertainers, leaders and led, teachers and disciples, navigators in outer space and toilers in the field, the rich and the destitute, the creators and destroyers, these and more, come from the same womb of humanity. People: "How noble in reason, how infinite in faculty! In form and moving, how express and admirable! In action how like an angel! In apprehension, how like a god! The beauty of the world! The paragon of animals!" as Hamlet exults.

These affirmations of our common humanity, the preciousness of life and our inescapable link with Planet Earth could breathe fresh life into education and help address some of the crying needs of our time, including the sense of alienation, despair, suicide, drug-abuse, homicide, and the desecration of our precious Planet resulting in unprecedented natural calamities and humanitarian consequences.

Obviously, engaging education to serve the wellbeing of the People and the Planet entails a collective, all-stakeholder effort beyond educators and educational institutions. Druk Gyalpo Jigme Singye Wangchuck of Bhutan envisioned Gross National Happiness signalling a bold departure from the conventional GDP-dominated worldview and presented a pathway for holistic and sustainable human and societal flourishing within mutually-supportive planetary boundaries. *My Green School* is a book I wrote that is a meditation on the core functions of education. It is seeks to restore education as the Noble Sector of public service and addresses the vital claims on the process of teaching and learning by empowering learners to cultivate the nobility of the mind, heart and hands, within the broad national vision.

MY GREEN SCHOOLS CORE FUNCTIONS

- Natural Greenery: our ability to discover and honour our vital links with all lifeforms around us and beyond us that sustain us
- Social Greenery: our ability to build relationships, goodwill and positive energy and releasing these to the society around us and beyond
- Cultural Greenery: the appreciation of who we are and what makes us who we are; our values, sensibilities, and worldview
- Intellectual Greenery: positive disposition to new ideas, knowledge, and information; openness of mind to seek and value new discoveries and insights and examine their merits
- Academic Greenery: our ability to discover and value the great ideas that define and give validity to the many academic disciplines that we study in our seats of learning

- Aesthetic Greenery: our ability to extend the range of our sensibility to appreciate objects and ideas that elevate and edify beyond the mundane and workaday
- Spiritual Greenery: an acceptance of the need for a higher, nobler and sublimer object to realise greater fullness and completion of our limited and unfulfilled lives
- Moral Greenery: our ability to distinguish between categories of values that give our special character as the human of the species

Green is a colour, but more importantly, it is a metaphor. Green symbolises anything and everything that supports and sustains life in its infinite variety – human, animal, plant, birds, reptiles – indeed all life and Planet Earth, above all. The global pandemic has been hard on us, but with all the agony and affliction that Covid-19 has unleashed on the world, there have also been some beautiful developments. Planet Earth has been able to breathe freer, the air is purer, water cleaner, vegetation greener, and birds and animals can reclaim their lost ground, even if for a little while. There has also been a spontaneous resurgence of the old latent seed of humanity in the positive energy and goodwill, the spirit of volunteerism, caring and sharing so manifest in all lands. I call this re-germination of basic human goodness Revolution 5.0: The Return of the Human. Covid-free, may these precious gains become the new norm before the old life-denying GDP-impulse returns and sucks out the soul of our Planet and humanity. We need to re-invest faith and hope and integrity in education and engage its vast potential in the renewal of vital relationships and the building of a world we all deserve.

Author of My Green School, *a vision of holistic education as an instrument for human and societal flourishing, Thakur S. Powdyel, believes in Education as the Noble Sector of public good that combines the grace of the head, heart, and hands to make a positive difference to the society and our* Planet Earth. *He remains engaged in public service and reflects on the outer and inner life of institutions and nations. Powdyel is the recipient, inter alia, of the Gusi Peace Prize International, and Global Education Award, for his outstanding life-time contribution to education.*

Harmonising Humankind with Nature

If education does not celebrate our connectedness and the wholeness and Holiness of it all, then what is education for?

David W. Orr

"Our ideas are too puny for our circumstances"
– Rev. William Barber

We are in the rapids of human history and capsizing is not a remote possibility. For those of us dedicated to educating youth for lives in a confusing and uncertain world the question is, 'What do we do?' Living in the shadows or the sunlight of our legacy, what would our great, great grandchildren wish us to have done?

They would likely ask us to see what is right before our eyes: heat, storms, fires, floods, desecrated lands, extinctions, and injustices and what these portend for their lives. They would ask us to reckon with the possibility that, to reiterate the quote from Reverend William Barber, "our ideas are too puny for our circumstances," and to think more wisely about what it means to be human. They would demand that we stop using the atmosphere as a dump and that we preserve Earth's forests, rivers, soils, seas, mountains, lifeforms, and grasslands. They would ask us to enlarge the democratic vista to include them, their great, great grandchildren, and other species. They would certainly want to live in a world in which words mean what they say.

As a source of such remedy, however, institutions of higher education are committed not to transformation but to patching the flaws in the deeply flawed modern paradigm on the wager that it carries the seeds of its own renewal. The educational system with millions of students each year, billions of dollars of research funding, trillions in capital assets, operates with the assurance that goes with its assumed monopoly of solutions to what ails modern societies. It exists unmolested in the world of influence and money as long as it does not threaten the prerogatives of capital and the underlying faith in economic growth and domination of Nature. Its very organisation impedes non-trivial conversations across disciplines. Its financial dependency limits serious reckoning with ideas of justice, peace, interdependence, and ecology. It deals in what E.F. Schumacher called. 'convergent problems' not 'divergent problems.' The former are linear and so amenable to scientific or technological solutions. The latter are more like dilemmas that are, by definition, unsolvable but avoidable with foresight. Increasingly our basic problems are of the latter sort, they are divergent moral and political questions, in Schumacher's words they are, "refractory to mere logic and discursive reason." For reflection or simply mulling such things over, the velocity of learning and research is too fast on some things, too slow on others. Too often, college and university graduates become what Wendell Berry once described as, "itinerant professional vandals."

A decent and durable future for humanity, however, will require agents of repair, healing, and transformation— practical visionaries with their feet on the ground and their eyes on a farther horizon. For them, the 'Great Work' ahead is to build a fair, peaceful, and resilient global civilisation powered by sunlight, while saving much of the Earth for other species, restoring degraded ecosystems, building an economy in which prices tell the truth — all undergirded by a deep reverence for life. Education that harmonises humankind with Nature starts with ideas contrary to those underlying conventional education:

- Ecological disorder reflects a prior disorder in the way we think and what we think about, making ecology central to all educators
- Humans are fast thinkers but slow learners
- The idea of 'systems' implying our connectedness with all that was, is and will be, is most radical and necessary in our language
- True self-interest is inclusive not exclusive, which is to say, 'I am because you are'
- Not all knowledge is good and not all of it can be deployed responsibly in a world of feedback loops, leads and lags, surprises, and long lapses between cause and effect; and new knowledge is not necessarily better than old knowledge rediscovered, ie. 'slow knowledge'
- Every act of analysis should be balanced by one of synthesis

- Formal education deals with half of the brain, ignores the other half and seldom engages the hands or heart. The result is often an 'inverted cripple' with a single overdeveloped analytic capacity
- Formal education, bounded as curriculum, can be completed in a few years but true learning is an unbounded process over a lifetime
- The important problems are those of education not those in education

On the periphery of formal education, many small educational centres around the world serve as important adjuncts to colleges and universities. They are not a substitute for formal education, but offer the opportunity for students, faculty, and others to step back and put things into perspective and to sort the important from the trivial. They function more like a compass that clarifies direction not like a map marking an itinerary.

Schumacher College in Devon, UK, is one example. The College occupies an old carriage house on an estate that dates back to 1388. Named after the author of *Small Is Beautiful*, Schumacher College concerns itself more with large questions than with answers. Typically, the questions posed in seminars and conversations at Schumacher are the divergent kind that challenge established paradigms and pomposity of any kind. The atmosphere is seldom as certain as in the higher reaches of the academic world. The scale is minuscule — several hundred students per year. Its clock-speed — the rate at which things happen — is human-scaled, i.e., the rate at which ideas can be comprehended and absorbed. Its stock in trade is the kind of dependable old knowledge that has accumulated over many centuries and in many cultures.

> Intimate engagement with people and place is rather like the effect of salt in stew: small by volume but large by effect

Daily routines at the College allow for serendipity and spontaneity. The focus is a kind of disciplined diversity and boundary-crossing thought. The program includes meditation, music, lectures, gardening, walks along the coastal fringe that embed us in 'deep time' geologic history. In other words, it is diverse but unified around the reciprocal connections of body, mind, and soul. The College clientele is diverse. The classes in which I have participated over the years included students of all ages and backgrounds from all over the world. Still, they typically bonded quickly into a supportive community in part because they work together to keep the place going. More important, at the periphery and removed from the mad bustle and busyness of their ordinary lives, participants have the time to sort the trivial from the important and observe the world and themselves from a calmer and saner vantage point.

I leave it to others to wrap these organisations and experiences into a proper pedagogy and philosophy and to envision what a planet-wide network of small life-centred teaching and listening institutions might do in the larger scheme of things. For students and facilitators alike, however, I know that intimate engagement with people and place is rather like the effect of salt in stew: small by volume but large by effect enhancing the flavour of the mysterious thing called 'education.' If we are to be truly drawn forth — the root meaning of the word education — we need such places and times to reconnect with our souls, the soil under our feet, and the Life all around us. And if education does not celebrate our connectedness and the wholeness and Holiness of it all, then what is education for?

David W. Orr is the Paul Sears Distinguished Professor of Environmental Studies and Politics at Oberlin College Emeritus and currently a Professor of Practice at Arizona State University.

Cultivating Safe Uncertainty

A Billion Portals of Possibility Have Opened in the Wake of the Pandemic

Jon Alexander

"If uncertainty is unacceptable to you, it turns into fear.
If it is perfectly acceptable, it turns into increased
aliveness, alertness, and creativity."
– Eckhart Tolle

Until the autumn of 2008, I thought I knew it all. I'd had the best education the world could offer, and I sailed through it – straight As, and a Cambridge degree. Along the way, I had seen the twin towers come down, and heard Bush, Blair and Giuliani tell us what we, the Western public, should do in response to defend our way of life and lead the world into the future: to go shopping. So, when I graduated, I went to work at one of the world's biggest advertising agencies, to help – the system worked for me, and I was keen to work for it. But it was at this point that I began, little by little, to raise my head and see the world around me – in particular, the already manifest urgency of climate change. Increasingly troubled, but still interpreting this as a bug in the system and not a major feature, I decided to go back to university to learn to help fix it. I found a course I could do while continuing to work, a Masters in Responsibility and Business Practice at the University of Bath. And then, at the first week-long gathering of my cohort, my brain exploded.

I remember the moment clearly. It was Friday afternoon, and I was in discussion with my tutor. We had spent the week talking, reflecting, telling each other stories of our lives and our dreams. Having signed up to learn how to fix the climate, I was frustrated. "I came here to learn the rules," I said. "That's what I do. I learn the rules, I play by them - and, generally, I win. I came here for you to teach me to do that with climate change. Not just to *talk*." I remember exactly what she said in response. "What if I told you there are no rules?" I stared at her. Then, from absolutely nowhere, I broke down in tears.

I believe this insight – the power of which I felt viscerally then and understand more fully now – must be the foundation stone of an education in which people and planet truly matter. It is not just that the rules we have been living by, that manifest in every structure and institution of our society, are broken, although of course they are – what we need to root our education system in, is the fact that *no one knows* what should replace them in order for seven or eight or ten billion of us to live sustainably and peacefully and joyfully on this planet. There is no point pretending anyone does; any prescription offered that has any certainty at its heart is a lie. We are living now in a time of radical uncertainty, and we have to start by facing that.

This might not seem a particularly fertile starting point for reimagining education because, *if no one knows, how can anyone teach?* To explain why it absolutely is, I want to draw on a little-known essay published nearly twenty years ago by family therapist Dr Barry Mason, called *Towards Positions of Safe Uncertainty*. In his essay, Mason argues that those coming for therapy tend to occupy one of two 'positions': *unsafe uncertainty*, characterised by anxiety and a loss of co-ordinates by which to navigate their lives; or *unsafe certainty*, characterised by self-disgust and rejection. All know what they think they want, and this Mason terms *safe certainty*: solutions, answers, fixes. The problem is not so much that this is wrong, but that it is a chimera: safe certainty simply does not exist; there are no such things, really, as solutions, or if there are, they should be understood as, "only dilemmas that are less of a dilemma than the dilemma one had."

As such, all that a therapist can really do is support the person to step into a position of safe uncertainty. "This position," Mason writes, "is not fixed. It is one which is always in a state of flow, and is consistent with the notion of a respectful, collaborative, evolving narrative, one which allows a context to emerge whereby new explanations can be placed alongside rather than in competition with the explanations that clients and therapists bring. A position of safe uncertainty is a framework for thinking about one's work, orientating one away from certainty to fit, a framework for helping people to fall out of love with the idea that solutions solve things."

Education, I believe, must seek to cultivate safe uncertainty in everyone. We need to see the challenges we face as radically unresolved in order to know that the only way forward is to see every single one of us as a participant in facing them. So, what would such an education system look like? It would not pretend that theories were final answers but would see information as equipment. It would see every young person as the potential source of the next advance in understanding, as power to be unleashed not controlled. It would see teachers as giants not in the fee-fi-fo-fum sense, but on whose shoulders young people might stand in their search. It would see teachers and students as side-by-side collaborators in a shared inquiry into the future, not transactional counterparts. And the work involved in creating it would be about renewal, not just resistance.

Head-on battles over content and context are exhausting the energy of so many in our education system today. This resistance – defending the arts, ensuring history is not taught solely through the eyes of the 'winners', protecting teachers' time and rights, ensuring children are fed and can learn at all – must continue, and will intensify because, as the dominant system collapses, those who have attained power within it will seek to do the same things they've always done, but harder. But shifting the mindset can and is happening in parallel; and it is generative, energising work.

In the UK, this kind of change is taking shape at the core of the education system, under the moniker of Big Education, Whole Education, or oracy (as opposed to just literacy and numeracy). These are narratives of education that are about building agency and power in young people, not just implanting knowledge; they are in many ways imperfect compromises but are growing communities of practice that meet the system where it is today, saying some of what needs to be said in a way that can be heard.

It is also coming in from the fringes, in the form of projects like the youth-led and -governed Rekindle School, a supplementary school in South Manchester where children and teachers are stepping outside the existing system to create the school they dream of in the time after hours. It is there in the resurgence of democratic- and Steiner-based education, in forest schools and more. Most of all, though, it is something individual teachers and community leaders and parents are simply stepping into on their own initiative and in their own daily practice, treating young people differently in small but vital ways, because they are thinking of them differently. It might lie beneath the scope of the headline writers. But it is there and growing. Especially now.

> Individual teachers, community leaders
> and parents are treating young people
> differently in small but vital ways, because
> they are thinking of them differently

At the outset of the pandemic, Arundhati Roy published an essay in the *Financial Times* in which she argued that "the pandemic is a portal"; that for all the tragedy of our time, a door to a different future had been opened. Today, some fear that portal is closing, with little having changed. I see it differently. That day in 2008 was seismic for me. But it would be two full years before I would leave the advertising industry, another four before I began in any recognisable way the very different work I now do. Seen from outside, I stayed in my previous world, reabsorbing myself into safe certainty. But inside, something had shifted. The external appearance was the same, but a portal had opened inside me, and it could only grow.

I believe that is analogous to where we are now collectively, as a species, not just in terms of education but in every aspect of our lives. Our certainty has been blown apart. It might look like the space is closing, but the opening is more subtle and distributed than Roy's metaphor would suggest. There is not one enormous portal, but billions of tiny portals inside our minds that are expanding and connecting at different rates. Many of the most vital and most promising exist in the realm of education, among young people who are less constrained by the world-as-is, and among those who work and live with them every day.

My hope in writing this essay is to help teachers and students and parents acknowledge their portals, the validity of their uncertainty, and feel safer in sustaining it. If reading it does that for even one, writing it will have been time well spent.

Jon Alexander is Founding Partner at the New Citizenship Project, and author of CITIZENS: Why the Key to Fixing Everything is All of Us.

Teach the Future

Lessons from the
Black Mountains College

Natalia Eernstman, Ben Rawlence and Tom Sperlinger

*"Other animals, in a constant and mostly
unmediated relation with their sensory surroundings,
think with the whole of their bodies."*
– David Abram

Before the pandemic quashed mass gatherings, young people worldwide flocked to the streets to express their dismay with the status quo. The school strikes didn't just call out grown-ups' inaction in the face of the climate emergency, pupils and students also walked out of classrooms to rebel against their education. They'd had enough of being locked in a system that doesn't prepare them for a future which is looking increasingly unviable. Education has always mattered, of course, but its role in reproducing a destructive economy – and its capacity to play a different, positive, role – elevates it to a zone of protest like never before.

Students around the world are calling on schools and universities to 'teach the future' -- and they must. But it is the moral responsibility of a proper education system not just to tell students the truth about the climate and ecological emergency but to prepare them for it. Putting that responsibility into practice is the animating idea behind an emergent institution on the border between England and Wales: Black Mountains College.

To confront and adapt to the unfolding planetary emergency we will need all the creative, adaptive, and collaborative capacity human beings can muster: competencies that mainstream education is currently not well suited to foster. Existing secondary and tertiary education is almost entirely based on attaining economic, rather than broader, more human goals: preparing pupils for a lifetime of work so that they contribute to the economy, while also supplying authorities with quantifiable data to hold schools and colleges accountable. The deteriorating mental health of young people which can partly be attributed to the increase in testing and pressure to achieve academically, shows that neither of these external goals contributes to wellbeing or the ability to live a fulfilling life. What's more, judging by the ecological crises, formal education is manifestly failing to teach people how to respect planetary life.

The case for a different, more complete understanding of education has been self-evident to reformers for more than a century. Educationalist John Dewey, for example, critiqued schools for being outdated institutions that prepare children for a world of the past. In fact, he said, school shouldn't even be about preparing pupils for future living – education should be the process of living itself, based in the real world. Education should serve human life, to discover and celebrate what it means to be human. It should also serve all life.

More recently, Gert Biesta described a pedagogy which focuses on democracy, ecology and care: teaching pupils to be in relation to the world without placing themselves solely at the centre of it. He calls this, "to exist in the world in a grown-up way" which is attained by developing the full potential of a person. This echoes the 'Buen Vivir' philosophy from South America that aims to educate humans to be in harmony with each other and their habitat, among others. Similar ideas have been the driving force behind educational experiments from Summerhill to Dartington, the Steiner movement, Reggio Emilia, Bauhaus, and Black Mountain College in North Carolina established in the 1930s. And yet, despite the proven successes of these so-called alternative approaches in fostering creativity, innovation, emotional intelligence and so on, mainstream pedagogy has remained stubbornly narrow, focusing on logo-centric activities that can be easily quantified and assessed.

In fact, as many people from Indigenous communities to educationalists like Maria Montessori and Loris Malaguzzi pointed out long ago, we know that this is a very partial view of how humans learn. As neuroscience is now demonstrating, humans learn with their whole bodies, activating all their senses. Memories are imprinted in all cells, not just brain cells. Neurological pathways are built by making links through experience. Humans learn through play, problem-solving, through engaging their emotions alongside their cognitive functions. None of this is new. And yet, the science of learning only seems to be taken seriously in early years education, rarely at secondary and almost never in universities. The planetary emergency demands that we urgently seize all that we know about how humans learn and apply it to the central challenge of our epoch: regaining our place within Nature.

> The science of learning only seems to be taken seriously in early years education, rarely at secondary and almost never in universities

The new tertiary programme at Black Mountains College (BMC) proposes that to best prepare for the challenges ahead we need to unleash the human: the core human competencies (that most employers, incidentally, also claim to want) of creativity, communication, care, collaboration, and critical thinking. With this in mind, BMC is designing a curriculum aimed at fostering skills (not subjects), based on three foundational principles: learning to learn; artistic and sensory training; and taught in Nature and applied settings of landscape and community.

To confront and adapt to a rapidly changing and uncertain environment, we need to be adept life-long learners. We need to understand how we learn individually as well as collectively in order to be responsive, resilient and resourceful members of communities that creatively confront challenges through improvisation and collaboration. The single degree programme at BMC therefore starts with asking the question 'how do we learn?' Students then take units blending the neuroscience of perception and cognition with artistic (visual, aural, movement) training to understand the foundations of human communication and interaction – with each other and with other lifeforms.

A transition to an ecological civilisation that respects planetary limits is of course, not a scientific problem but a problem of behaviour, culture, and politics. And the currency of behaviour, and politics, as we know, is not argument but emotion, experience and the imagined collective. Sensory training and the practice of the arts is therefore essential pedagogical territory for an education that not only teaches but prepares for the future. Importantly, this is not about the theory of the arts or the logo-centric rhetoric of reading and writing, but creative practice: ways of seeing, hearing and moving; ways of creating, expressing and communicating.

The practice of the arts is common to all human societies. It is a democratic space for exploring different traditions, cultures, and worldviews. It acknowledges different ways of seeing and knowing. Creative practice also teaches a method for approaching problems that is rooted in improvisation, play and experimentation. As an evolutionary human attribute, play teaches us to collaborate and co-exist, it reduces anxiety and stress, and helps us to creatively learn our way out of adverse conditions. Knowing how to play and improvise, and continuing to do so as an adult, thus forms a key survival strategy for the Anthropocene.

Acquiring skill in a sensory field invites a radical re-appraisal of the meaning of the term 'interdisciplinary' that goes beyond the traditional blending of western canons of thought, the arbitrary 'disciplines' such as politics, philosophy and economics, examined only through a single cognitive lens. It widens the definition to encompass all spheres of human expression and participation in the world. This leads to the third principle of BMC: learning in the 'real world' on applied problems for meaningful and socially useful ends.

Learning to live in the Anthropocene means learning to regain our place within Nature, learning to live within planetary limits. While this may sound simple, the gap between current resource use and safe planetary limits is huge. Mainstreaming ecology is, belatedly, becoming a fad in educational discourse, but learning about Earth systems is not enough. Our current crisis is not due to a lack of environmental science. What is missing is a critical awareness of the limitations and potential of human systems. We are prisoners of what we imagine to be possible. We need leaders who see the world from an ecological starting point, who appreciate the scale of the challenge ahead, and who understand how emotion, perception and action combine to change behaviour.

How do you inculcate such an ecological perspective that accords non-human life appropriate respect? The first step is by making Nature itself the classroom. BMC is sited on a 120-acre farm and surrounding mountains. Sensory training outside will help to embed and embody the ideas of Merleau-Ponty updated by David Abram that 'all perception is participation' to open the door to seeing oneself as part of the biosphere, not separate from it.

Science has shown Dewey's ideas to be correct: the most durable lessons are those derived from problem-solving in context, in applied settings. This is reflected in the third year at BMC designed to map the students' way in the world with a work placement and a student-driven research project keyed into realising the ambition of the unique Welsh law – the Well-Being of Future Generations Act – together with a seminar on theories of change examining where and how students might best apply their own talents. The purpose is to demonstrate and interrogate the tangible change that is possible so that graduates of Black Mountains College are as well-equipped as possible to answer the eternal question facing us all: what is the right thing for me to do, for myself, society and the planet, at this particular moment in time?

Dr Natalia Eernstman is an artist, researcher and educator who works across the Arts and Sciences to build community resilience in the face of a climate-changed future. She leads the MA Creative Education at Plymouth College of Art.

Ben Rawlence is the founder of Black Mountains College and the author of three books, most recently, The Treeline: The Last Forest *and* The Future of Life on Earth.

Tom Sperlinger is a Professor at the University of Bristol and Academic Lead for Black Mountains College; he is co-author of Who Are Universities For?

Environment as Educator

TAKING INSPIRATION FROM THE LONG AND PROVEN HISTORY OF OUTDOOR LEARNING

Elizabeth Howes

"Of all the paths you take in life,
make sure a few of them are dirt."
– John Muir

When I was 17, I decided to be a primary teacher. I had always cared about the environment and the natural world, and teaching seemed an excellent way to pass this passion on to others. Growing up with a father who worked for Parks Victoria and a mother who came from the country, and whose father was a high school principal renowned for his deep concern for students' welfare (he designated a study day every week where students could learn at their own pace and in their own homes), I was perhaps destined to combine the two professions. I am also someone who is interested in too many things to specialise – Nature, languages, reading, music, maths, history, and dancing, to name a few – and being a generalist primary teacher seemed an ideal way to explore, and help children to discover, these different passions.

With this dream in mind, I left school and embarked on a gap year at an environmental education centre in the Lake District called Castle Head. Located in one of the most picturesque parts of the United Kingdom, this centre hosted school camps, ornithology conferences and corporate team-building events. My job was to assist the outdoor educators with the activities: canoeing, rafting, rock climbing, hiking, and the 'high all-aboard' – a tiny platform at the top of a towering pole, on which a group of four people had to collaborate to balance.

It became apparent very quickly that the visiting city children were seriously disconnected from Nature. Many were unused to walking any distance; many had never seen open space or farm animals in real life. One London teenager expressed surprise that sheep, "don't look like clouds, like I thought – they look like mops," while another was appalled that when relieving oneself in the outdoors on a cold winter's day, the difference between your body temperature and the ground temperature will cause steam. Gradually, these children managed to stretch their comfort zone and adapt to the outdoors. We received many letters of thanks for the experiences and the new skills and character traits they had discovered. I hope that some were inspired to seek Nature for themselves in later life.

Witnessing these children's initial disconnection from Nature, and their gradually increasing comfort with it, confirmed my dream of pursuing environmental education. I have always believed that taking care of the environment should be humanity's first priority: if we do not have a habitable planet, we do not have anything.

I returned from the UK to Australia, completed my Primary Education degree and got a teaching job in a local school. I soon realised that, while environmental sustainability is briefly mentioned in the curriculum, the push for learning through technology and the pressures of NAPLAN (Australia's standardised testing) means that environmental sustainability is often pushed down to a very low priority and taught in the most tokenistic of ways. But my dream of focusing on environmental education persisted. There are similar environmental education centres in Australia to Castle Head, and I considered seeking work in that field. However, I believe strongly that a classroom teacher has much more leverage and the ability to bring about lasting change. I determined to do what I could in my role, starting a lunchtime gardening club and an unofficial Farmers' Market, where the children sold produce from the school garden for donations to the school community.

In 2018, I discovered an opportunity to study a Forest School Leader course. This rekindled my hope that I could bring more environmental education into the classroom. I signed up for the course, and it shaped the course of my professional life.

At its simplest, Forest School is child-led, play-based, outdoor learning. It has deep roots: free, outdoor play as a means of education is innate in humans and has been since the time of hunter-gatherer societies. The psychologist Peter Gray, who defines education as 'cultural transmission' notes the way that hunter-gatherer children, "play at, and therefore practice, all the activities that are crucial to the life of the band." Such activities include identifying and mimicking animals' sounds and habits, finding secret hidden places, and building shelters and huts, making and using tools for digging, hammering and drawing. All these activities, and many more, make an appearance in Forest Schools. The Forest School pedagogy aligns with educational philosophies such as Montessori, which emphasises long, uninterrupted blocks of time for child-led learning and the use of real tools; Steiner, which places an importance on learning about the natural world; and Reggio Emilia, which involves self-directed, experiential learning with the environment as educator.

I have been leading a Forest School program at a local school for a year now, and it is extremely heartening to see the myriad of benefits it offers the children. These can be summed up simply as: connection to nature; connection to their innate desire and drive to learn; connection to each other.

CONNECTION TO NATURE: David Sobel, inventor of the term 'place-based education,' has shown that environmental values and behaviours in adults correlate directly with, "childhood participation in 'wild' Nature" which Forest School provides generously. For children, local, hands-on, down-and-dirty environmental experiences create a far more lasting connection to Nature than learning information about global environmental issues, which can paradoxically create a sense of fear and helplessness. In today's globalised world, information about the global environment is easily available and, in light of the COVID-19 pandemic, digital technology has been seen as a saviour and a success, particularly in the context of remote education.

However, the pandemic has also highlighted the importance of hyperlocal connections, especially during lockdowns, travel restrictions and 5-kilometre radius limits. Many families in Melbourne found themselves and their children becoming intimately acquainted with local pockets of Nature in their area – a definite silver lining and a reminder that, in a world of increasing digital technology and globalisation, connection to one's own natural backyard is what ignites the spark to protect the natural world. Louise Chawla of the University of Colorado found that two predictors of environmental activism in adulthood were, "many hours spent outdoors in a keenly-remembered wild or semi-wild place in childhood and adolescence, and an adult who taught respect for Nature."

CONNECTION TO THE INNATE DESIRE TO LEARN: Peter Gray has described the history of the current mainstream education system, from religious institutions seeking control and seeing children as inherently sinful, to the time of the Industrial Revolution, when the goal was making workers rather than guiding individuals. Although this aim has evolved and shifted over time, the system remains much the same. The idea of student voice and agency, for example, receives a great deal of lip service these days, but is incompatible with the pressures of a standardised curriculum and standardised testing, which results in teachers being forced to 'teach to the test.' Forest School, on the other hand, not only fosters connection to and knowledge about Nature, it also gives children genuine empowerment and agency over their learning. Nature is a great leveller – everyone can participate and learn, whatever their interests, strengths or entry point. Differentiation, therefore (another buzzword in modern education) becomes a *fait accompli* – every child can work at their level in an outdoor, play-based setting. There is no standardised testing in Nature.

There is no standardised testing in Nature

CONNECTION TO EACH OTHER: Today's education system has its origins in rote learning and the memorisation of knowledge, and vestiges of this remain in the standardised curricula of so many countries. In a world where information is so easily accessible, this is not the most important aspect of education. Instead, critical thinking, creative thinking, problem-solving, social and emotional intelligence and resilience, and emotional intelligence and resilience, are all vital skills. Forest School provides an ideal environment for children to learn these skills. There is less time pressure, so if a child needs more time to work out a solution to a challenge, this is available. The long blocks of time allow for multiple attempts at a task with modifications, which lead to resilience – an essential skill for the challenges of life. When children work out the best way to climb a tree, or work with other children to find the best fallen logs to build a shelter, they are assessing, attempting, inquiring, experimenting, refining, risk-taking, negotiating and collaborating – all of which are vital skills for social intelligence and for the whole of life.

My vision, therefore, is that every school will adopt a Forest School program as a matter of the highest priority. An oft-used quote, paraphrased from Wendell Berry, reminds us that, "We do not inherit the Earth from our ancestors; we borrow it from our children." What better gift, then, to our children, than to nurture their connection to Nature and empower their determination to protect the world they love?

Elizabeth Howes is a primary teacher and Forest School Leader based in Melbourne, Australia. Her passion is to help all children rediscover their innate connection to Nature.

The Sum of Solutions

Utilising Problem-Based Learning to Solve Critical Problems

Michael Keary

"Once children learn how to learn, nothing is going to narrow their minds. The essence of teaching is to make learning contagious, to have one idea spark another."
– Marva Collins

One concept unites both environment and education: waste. We waste water, food, energy, and irreplaceable ecosystems. We also waste time, effort, opportunities, and invaluable young minds. In our ways of living, working, and educating, we must become far more efficient. We can start by remodelling our education system. Rote learning of abstract content is still the dominant mode of learning globally. The ineffectiveness and somnolence of this approach is bad enough, but it also misses the opportunity to utilise learning in the world's efforts to solve critical problems.

Instead, we must embed problem-based learning (PBL) at the heart of education. In a problem-based learning approach, everything a student learns helps them solve a problem. Students no longer learn subjects passively but by applying the provided concepts and skills to relevant, real-world challenges. Both real-world relevance and active learning super-charge engagement. So does the collaborative nature of PBL, which also enhances team-working and general social skills. With a problem-based learning approach, we can reduce boredom, reuse tired content innovatively, and recycle countless wasted classroom hours.

> Both real-world relevance and
> active learning super-charge engagement

Above all, we can use this precious time to develop better citizens. Imagine a school with a holistic curriculum: where physics, chemistry, economics, history, philosophy, literature, biology, politics, and mathematics are interdisciplinary, treated as complimentary routes into understanding the globe's most critical problems. Amongst these problems is climate chaos, the greatest challenge humanity has ever faced.

At this holistic school, on Monday morning, 9am, a student learns about covalent bonding in carbon dioxide; by 10am, she is discovering metre through the work of green poet Nicanor Parra; she has debated the role of the state before lunch; and will walk home with her mind full of linear equations for methane emissions. Nothing is lost with this curriculum – indeed traditional content knowledge is communicated far more effectively. But think of what is gained! Every class presents students with a problem, each of which feeds into a greater problem. The law of conservation of energy is learned through examining energy storage issues, but as part of a physics course focused on the science of climate change. Nothing is abstract anymore. Students know that all their learning has a meaningful end: equipping them to be the kind of citizens we need. Citizens with the educational foundation for any future that appeals to them but equipped also to participate in the political process of dealing with the environmental crisis across its many dimensions.

Some of these dimensions involve class. How are we to redistribute wealth so that wins for the environment are not defeats for equality? Welcome to your economics class. Others involve prejudice: to address climate change, we need a stable society in which citizens work together. How are we to achieve this if substantial sections of our society suffer from institutions and norms that are fundamentally racist? Sociology will begin shortly. The environmental crisis is a wicked problem: its solution is really the sum of the solutions of many near-equally tricky problems.

So, what is stopping us from implementing this ecologically and socially focused, problem-based learning approach to education? Pedagogically, PBL is more or less universally acknowledged as vastly superior to the passive, abstract learning approach that has heretofore dominated our schools and universities. It has, moreover, been implemented in some form in many schools and universities. The first issue is inertia.

Too many experienced teachers are hostile to novel pedagogical methods, having grown comfortable with their own; and too many inexperienced ones want their classrooms to resemble those in which they themselves succeeded. Also, many of PBL's insights are counter-intuitive: lecturing is ineffective; the teacher should not be the centre of attention; talking in class is to be encouraged; students should own their own learning. As a result, though many pedagogical professionals are at least aware of PBL's existence, they have no desire to understand it further, let alone implement it. Finally, assessment drives learning, and too many our assessments were designed decades ago. They thus assess what a successful mind was thought then to encompass: accumulated facts, abstract formulas, and a base level of intelligence. These assessments are alien to the notion of education as skills one must apply. Continuing professional development for teachers must have PBL and other active learning strategies at the heart of their design if these barriers are to be overcome.

The second issue is awareness. Incredible though it may seem, too few people in positions of power in education have more than a vague grasp of the extent and causes of the environmental crises. This is largely because their exposure to the crises has been through a media and political establishment that is wedded to capitalism. In this establishment, the problem is technical. To solve it, markets must be put in place, technologies developed and dispersed, bureaucracies built, and investment capital gathered. That the environmental crisis is a consequence of a consumerist, individualistic citizenship, underpinning an ecologically rapacious globalisation, is not an idea much disseminated outside of academia and social movements. As a result, educators feel their role is merely to boost the number of STEM students, rather than to facilitate the development of ecologically aware, critical, and independent learners.

Only by growing social movements, and hence political and media awareness, can the powerful in education be brought to see the scale of their responsibility. If this can be accomplished, problems beyond the environmental will be eased. One such is inequality. Better teaching helps poor students more than it does rich ones. Better-off students can turn to parents for help, parents who may be educated enough to provide tutoring themselves or who can afford to pay for afterschool help. The less well-off rely much more on their schools. More effective pedagogical methods mean more students achieving learning-outcomes during school hours, dissolving much of the advantage provided by afterschool help.

Moreover, engagement comes on average less easily to worse-off students. A student whose single parent works full time may lack the disciplinary environment often crucial to motivation. Providing more engaging lessons also smooths out this gap.

There is rightly much focus on the infrastructural needs of modern education; the equality gains that could be made with more funding for facilities in poorer schools. However, according to the OECD, significant additional spending over the past two decades has made no discernible difference to student learning outcomes. Yet changing our delivery mechanisms, a reform that demands few, if any, additional resources, can be transformative. We should not lose sight of the fact that schools are wasting so much of what they already have.

The world is beset with problems, the most pressing of which are the environmental crises. And yet our politics is becoming less, not more, able to deal with problems of great complexity. Populists, pedalling easy answers and sowing division, are perceived as the most trustworthy. Many citizens are clearly ill-equipped to assess reality accurately. We have no better tool to combat this than our education system. Yet we waste it on out-dated teaching methods and abstract content and ignore the pressing need for graduates who can pursue good evidence, evaluate alternatives, and create novel solutions. Let us instead put our education system to work on the problems affecting our people and our planet and help it deliver the citizens we need to solve them.

Dr. Michael Keary is a Global Politics Tutor at King's College London and a Fellow of the Higher Education Academy. His research, which lies at the intersection of environmental politics and political theory, explores how perceptions of technological change shape the treatment of the environment in both theory and policy.

Lessons in Languaging

WHAT MATTERS TO YOU, EARTHLING?

Angela Dawn Kaufman

"Imagination is more important than knowledge.
Knowledge is limited. Imagination encircles the world."
– Albert Einstein

What is today's learning objective? You tell me at the end of the lesson. All you need to know for the moment is that this is an interdisciplinary lesson, where no subject takes precedence, but all are interconnected, interdependent, and integrated, and all you need is comfy clothes. Are you ready?

TASK ONE: Close your eyes, wherever you are. Imagine a world that is upside down, inside out and back to front. Imagine that for just one lesson, nothing matters. If you do that, what will you unearth? You are a person; you have a mind, a heart, a memory. You think, you love, you remember, you feel. You also live on a planet. You are not alone. That planet goes by the name Earth because that is what it is made of. So, what matters to you, Earthling?

TASK TWO: Get up from your chair and head for the door that opens out to an open space and fresh air. Perhaps you can walk on grass, stay in the shade of a tree, or hear water running. As you walk, be aware of the ground below you. How does it feel to be in your shoes? "You are like an acrobat rolling an enormous ball – the great round Earth," suggests Stephan Harding in his book *Animate Earth*. How does it feel to be the ground below the soles of your feet? What does the Earth feel as you take each step? Does she feel the weight of you as you move forward to your destination? Do you help keep her in motion as she rotates and orbits the sun, or do you tire her? You and almost eight billion other pairs of feet?

TASK THREE: Look around you and find a patch of Earth. A bush, a shrub, a tree. Without pulling up any roots of plants, take a small handful of that earth, that soil. And now, really look at it. What colour is it? Is it damp or dry and does it smell? Can you see stones, dead leaves, or bits of rubbish? Can you squeeze or shake it? Does it make a noise?

TASK FOUR: Think of the many souls who have walked here before you in this place. Perhaps it has changed over and over, like a chameleon or an octopus, unrecognisable even to itself. Or perhaps not much has changed at all from the day your ancestors still breathed. So, imagine you can see them now, those ancestors. You can meet them, look them in the eye and ask them a question. What would you ask? What would they say to you? Joanna Macy, in her book *Active Hope* describes the possibility that, "if our intention is clear, we can travel back a century to enter the hearts and minds of past beings." Would you apologise or would you expect them to apologise to you for their mistakes?

TASK FIVE: Find two trees, some distance from each other. (If you cannot see any trees near you, then in the next lesson we will find a tree-planting project, of which there are many, or create our own.) One tree will be the start and finish line, the other will be your half-way mark. Flex your muscles, stretch, and begin breathing with deep breaths because you are going to run, or walk briskly. Just keep up the momentum. As you run the laps forward and back between the trees think about the millions of people who must walk for hours every day to collect fresh water in order to feed and wash themselves and their families. Keep moving. Later you can work out the actual distance you travel and compare it to areas of poverty in the world where keeping fit is not a luxury.

Sit down now and while you rest, look at your hands. Are they still grubby with soil from Task Three? Imagine the hands of children in the mineral mines of Africa, scrabbling for the coltan for your smartphone. Are they as dirty as that?

Breathe deeply. Close your eyes again and listen to your heartbeat. Feel it pump your blood around your body, blood that is oxygenated by that last breath you took. Enjoy the tiredness in your muscles, massage a cramp, feel the beads of sweat tingle from you. Smell them, taste them. You are alive, aren't you?

TASK SIX: As you regain your breath and energy, you are going to time-travel again, this time forward into the great unknown. Imagine your future great-grandchildren. Who are they? What do they look like? Do they look like you? Write them a letter. Think about what you will tell them about who you are, your life here and now, your work, your dreams, your goals, your frustrations, your advice. Tell them about your friends and family, your home, your school, your world. What will they understand of you in a hundred years?

Look around you. Where are you? What can you see? Draw a picture to help them imagine this place with you. What will they understand of this world today, then? Joanna Macy reminds us, "Those living in the future will look back on us as their ancestors. Recognising them as our kin brings them closer to us. A sense of care and responsibility arises naturally." Can you feel it?

TASK SEVEN: On your own, remember a moment during this lesson that caught your attention, resonated with you or that you found difficult. Focus on that moment and now imagine you are holding it in your hand. Look at it closely and remember how that moment made you feel. Cup it in your hands and breathe. When you open your hands, that moment is a verb. A doing word. Not a thing but an action. Explore in movement how that moment moves. How it moves you. Experiment with your own body what it means 'to ancestor', 'to poverty', 'to earth of dead leaves'. "We all take our first dance steps in the womb," says Donata Zocca. Remember this if you are having trouble letting go. Try to let go.

Later, in a group, bring those moments together and find a way of connecting them. Use the space, use a sequence, accompany them to music or sounds. Make the music and sounds yourselves. Let everyone contribute, talk about and ask each other what, how, who, when, where, and why. But above all, *do*, even in stillness. "Everything we do is a structural dance in the choreography of coexistence," said Francisco Varela and Humberto Maturana in their book, *The Tree of Knowledge*, as they contemplated how we gain knowledge of knowing totally through our body, whether concrete or abstract. As you think through your movements, and you move through your thoughts, know this. You are 'languaging'. While the word may sound strange, and my computer doesn't even recognise the word, it is one of the most simple, most natural and most human activities you will do today. So, enjoy it.

Make it *matter*.

All of the above can be extended, performed, filmed, edited, documented, repeated and varied depending on your circumstances and conditions. The activities can be shared digitally and remotely, even sponsored for charity or simply be part of a fleeting afternoon experience in the garden. In an 'Upside Down World' the teachers can be learners and the learners the teachers. Learning and teaching here are not separate entities. And what if the teachers aren't even human but animals, insects, trees, water, the sun? The real "fundamental facts of life," according to Fritjof Capra (in *Smart by Nature*), emerge from the diverse patterns that we see in Nature; networks, nested systems, cycles, flows, unfolding spirals of development and feedback loops. Can we not feel these when we dance? "In a world as interconnected as ours, protection of self and protection of planet are indistinguishable. If we act on our ability to mimic life's genius, we have a chance to protect both," suggests Janine Benyus. We need a "quieting of the voices of our own cleverness" while we listen to and through our only and closest tool – *ourselves* – because "The world *is* my body."

What if the teachers aren't even human
but animals, insects, trees, water, the sun?

Why should we have forgotten that? So, what is the learning objective of Lesson 2022? Perhaps you, the reader, have already answered that for yourself? Perhaps you already have new ideas to bring forward? I hope so, and I leave you with a new 'task'.

Logistically we may not be able to go back in time, but we can learn from past mistakes, and we can make a conscious choice to reshape and question – if not better, then differently, with whole new languages and by 'languaging'. Whatever that means. The place of education seems like a good place to start finding out.

TASK 2022: Question – "When is the best time to plant a tree?"
Answer – "Twenty years ago."
Now dance that.

Originally from London and trained as a classical ballet dancer at the Royal Academy of Dance, Angela has lived and worked for over thirty years in Northern Italy as a teacher of language and dance in many different spheres, and at present is looking to create new pathways towards 'being the learning' – with a universal vision and accessible to all.

From Conquest to Participation

Reimagining Education as a Living System within a Community of Practice

Lauren Elizabeth Clare

"Building community is to the collective as spiritual practice is to the individual."
– Grace Lee Boggs

In an increasingly unstable and rapidly changing world, there can be no higher value or greater measure of success than the cultivation of wellbeing within disruption. With the global decline in social, spiritual, and ecological wellbeing it is imperative that we learn how to counter volatility with vision, meet uncertainty with understanding, react to complexity with clarity, and remedy ambiguity with agility. However, the cultivation of these qualities is not a practice within current education systems. Rather, education has become a contributing factor to disruption, generating graduates who lack a deep understanding of self, community, and Nature – and the healthy capabilities that arise from these relationships.

Vital to human development and cultural health are experiential learning, inquiry, and the practice of emergent learning, yet these are not part of current education systems, which focus primarily on relaying trending facts and skills. The symptoms of escalating teacher and student burnout, fragmentation of mental health, and increasing social, spiritual, and ecological divisions all show that a different approach to education is needed. While education systems are an accomplishment of complex networks that reach around the world, connecting individuals, organisations, cultures, philosophies and perspectives, their detrimental symptoms show that the education tools used by these networks are helpful only on the surface.

While alternative education movements worldwide are demanding a new capacity in education and strive for holistic, humanitarian learning, these movements often fail to move beyond protest and dissent towards becoming influential forces that are truly regenerative. Without a conscientious foundation, educational experiences cannot encourage resilience, bring awareness, or pass on the inheritance of a culture. The regeneration of education and learning requires a shift not in the structure of our education system but the very foundation of our interactions. We need to seek the source of learning itself.

Disruption is not new. In every age, humanity has dealt with change by upgrading towards greater wellbeing and building a new collective experience. It is important to remember that we already have what we need in order to rise and meet our current challenges – within each of us is the inherent ability to learn from all of life, and to learn from disruption itself. We are now in the midst of an awesome humanitarian shift because the fragmentation of our relationship with Nature, self, and society is culminating as the end of the 'Age of Conquest.' However, the prescriptive foundation to this dying concept is still practiced within current education systems through the domination of facts and skills. While this model served a purpose as a step toward what we call 'progress,' its human-centric endeavours have endorsed a worldview that has led to the devastation of countless living systems for people and planet. Witnessing the destruction of ecology, psychology, and community brings the stark recognition that human beings are living systems, have evolved in living systems, and progress in living systems. Therefore, education needs to reflect these living systems and cultivate the human potential of learners, engaging our multifaceted nature, and awakening our latent powers of innovation, imagination, creativity, and wonder. Contrary to current educational reform movements, it doesn't take massive resources to engage change to a living systems view of education.

A living systems foundational approach that accesses the source of learning is not something that needs to be invented. We can transform the current education system from the inside out just with a shift in perspective of our interactions: from that of conquest to that of participation. Not only is this possible but it has always been accomplished through an approach currently known as the community of practice. Community of practice is an effective, robust, and time-honoured approach of participatory learning that creates the conditions to support holistic education and humanitarian development. Since the dawn of humanity, the community of practice has been the inherent model for how we come together in ethical space to learn from one another, from our environment, and from all of life. As a social structure requiring conscious participation and co-creation, the highly intentional community of practice is becoming known as a 'generative community,' for incorporating the tools and practices of mindfulness and wellbeing and implements learning for both individual and collective development. While the participants may engage in the collaboration of dialogue and activities, it is the highly intentional practices of deep listening, generative conversation, and co-creation that allow for discoveries to emerge from the synergy of individual, collective, and ecosystem. The co-creative experience allows for a gestalt shift in the essential understanding of interdependence and facilitates the recognition of human development as an aspect of the unfolding of one's ecosystem, intricately interwoven with the living systems of which one is a part. This is how the sustainable traditions of ancient social structures evolved to deal with disruption – by intentionally participating in the evolution of consciousness. These wisdom traditions show us that we do not need to 'unlearn' or 're-learn' or even 'learn how to learn'. The source of learning is to allow, invite, and hold space for the inner integration of knowledge and experience.

We can transform the current education system from the inside out just with a shift in perspective of our interactions: from that of conquest to that of participation. The further recognition of living systems is essential if we are to move beyond the protest and dissent of 'alternative education' to learning practices that are truly regenerative and serve the wellbeing of all of life. As Otto Scharmer, author of *Leading from the Emerging Future* reminds us, "We cannot change a system unless we change consciousness. And we cannot change consciousness unless the system can sense and see itself." Within the experience of participatory learning in a highly intentional community of practice we come to understand how the activity of the individual and the collective support one another to generate a living system. The American systems scientist, Peter Senge, encourages us to see that people are agents, able to act upon the structures and systems of which they are a part. This ego-to-eco perspective is, "A shift of mind from seeing parts to seeing wholes, from seeing people as helpless reactors to seeing them as active participants in shaping their reality, from reacting to the present to creating the future."

The symbiotic and regenerative balance achieved by the generative community as a living system manages negative internal impacts while allowing for the growth of both the individual and the collective; uplifting the individual to overcome the oppressions of groupthink consensus or the isolation of circumstances and honouring the collective wisdom by taming the ego of the individual. This dynamic generates a unique form of psychological safety which then supports the pursuit of inquiry/co-inquiry, reflective practices, and constructive feedback. It is conducive to the synergy of indigenous forms of learning, wisdom traditions, and the diversity of social innovation and ethical space needed to evolve the education system. Most importantly, this container allows for the quiet moment of integration that is the source of learning itself. The community of practice is then the key to cultivating participatory learning and encouraging regenerative practices for individual, collective, and ecological wellbeing.

An innovation to bridge the divide from current dominant education models to that of participatory learning can be accomplished with teaming practices which encourage creativity and collaboration. Another key innovation is in assessments, which provide an important reflective practice for growing individual and collective awareness. Designed with the recognition of supporting holistic agency rather than testing for accuracy of specialisation, it is possible for assessments to exhibit progress in curiosity, compassion, and courage, and can be documented through a portfolio of experiences. A synergy of self-assessments, peer-assessments, as well as educator and community assessments can support us in avoiding outdated conquest pedagogies, or the lure of providing progress for corporate culture.

Emerging examples of such innovative and generative communities of practice are The Presencing Institute – cultivating awareness-based systems change; Turtle Island Institute – cultivating indigenous social innovation; Dartington Trust and Schumacher College – cultivating regenerative and transformative learning practices; Shelburne Farms – cultivating learning for sustainability; Extinction Rebellion – non-violent civil disobedience for Climate and Ecological Emergency and regeneration; Weston A Price Foundation – nutrition advocacy and traditional foods education; The Society for Organisational Learning – cultivating capacity for organisational learning; Permaculture –

cultivating regenerative agriculture practices, and there are many more.

The regeneration of education and learning begins with the cultivation of wellbeing within disruption and continues by bridging the divides of ignorance. The current challenges of failing ecosystems, fragmenting mental health, and faltering social systems offer the opportunity to remedy the social disease of indifference and educate for greater social, spiritual, and ecological wellbeing. The conscientious innovations of highly intentional, generative communities of practice are rising to meet these challenges by cultivating participatory learning practices and the ethical space that supports psychological safety, nurtures wellbeing, and allows deeper sources of knowing to arise and integrate. They are demonstrating that the evolution of education and learning is to activate a shift in perspective from valuing domination to appreciating participation. For the wellbeing of people and planet, it is now essential that we awaken each other to our latent powers of innovation, imagination, creativity, and wonder.

 *Lauren Elizabeth Clare is the co-founder and facilitator of Regen Collective, a community of practice that was developed in the programmes of The Presencing Institute to support educators as a regenerative and evolutionary force for humanity. Lauren brings together formal and postformal educators in many disciplines from around the world to explore participatory learning, develop awareness practices, and collaborate through generative projects. Essays, reflections, and project developments can be read at Medium.com.*

GLOBAL RESILIENCE PUBLISHING
and PIPPA RANN BOOKS & MEDIA
imprints of
SALT DESERT MEDIA GROUP LTD., U.K.
Working in collaboration with international distributors, including Penguin Random House India.

Salt Desert Media Group Ltd (est. 2019) is a member of the Independent Publishers Guild. At present, the company has two imprints, **Pippa Rann Books & Media (PRBM)** and **Global Resilience Publishing (GRP).**

GRP began operations in Autumn 2021, with the first publications are planned for release in 2022. As the name suggests, the imprint focuses on subjects such as:

- Climate Change,
- The Global Financial System,
- Multilateral Governance (e.g., the United Nations),
- Public-Private Partnership,
- Leadership around the World
- International System Change,
- International Corporate Governance,
- Family Firms around the World
- Global Values,
- Global Philanthropy,
- Commercial Sponsorship, and
- New Technologies including Artificial Intelligence.

Two things make GRP unique as an imprint:

1. Our books take a global perspective (not the perspective of a particular nation);
2. GRP focuses exclusively on such global challenges.

By contrast, **Pippa Rann Books & Media (PRBM)**, launched on August the 17th, 2020, publishes books by people of Indian origin on any subject, but focuses on India, and on the Indian diaspora. Please see the list of books already published, and the books forthcoming from PRBM, at www.pipparannbooks.com

Both GRP and PRBM are open to first class ideas for books, provided complete manuscripts can be turned in on time.

Please note that GRP and PRBM exclusively publish material that nurtures the values of democracy, justice, liberty, equality, and fraternity.

***

Global Resilience Publishing and **Pippa Rann Books & Media** are only two of several imprints that are conceived of, and will be launched, God willing, by Salt Desert Media Group Ltd., U. K. The imprints will cover different regions of the globe, different themes, and so on. And if you have an idea for a new imprint that you would like to establish, please get in touch.

Prabhu Guptara, the Publisher of Salt Desert Media Group, says, "For all our imprints, and for the attainment of our incredibly high vision, we need your support.

Whatever your gifts and abilities, you are welcome to support us with the most precious gift of your time. The service you do is not for us but for the sake of the world as a whole. Please email me with your email, location, and phone contact details, letting me know what you feel you can do. Could you be an organiser or greeter at our events? Could you ring people on our behalf? Write to people? Write guest blogs or articles? Write a regular column? Do interviews? Help with electronic media, social media, or general marketing? Connect me with people you know who might be willing to help in some way or other?"

He adds, "I am one man working from his dining table, so I do not and cannot keep up with everything that is happening. There are many challenges and numerous opportunities – help me to understand what these are. Pass information on to me that could be useful to me. Put your ideas to me. Any and all insights from you are most welcome, as they will multiply our joint effectiveness. It is only as we work together that we can contribute effectively to changing our nation and our world for the better".

***

Join our mailing list to discover books from Global Resilience Publishing which will inform you on a wide range of topics, inspire you, and equip you as an individual, as a member of your family, and as someone who wants to make the world a better place.

www.globalresiliencepublishing.com

Printed in Poland
by Amazon Fulfillment
Poland Sp. z o.o., Wrocław
15 July 2023

373b6fb8-4580-4519-9c04-e18bdf8558d5R01